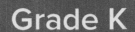
Grade K

Reveal MATH®

Differentiation Resource Book

mheducation.com/prek–12

8787 Orion Place
Columbus, OH 43240

ISBN: 978-1-26-421066-4
MHID: 1-26-421066-3

Printed in the United States of America.

Mc
Graw
Hill

mheducation.com/prek-12

Send all inquiries to:
McGraw-Hill Education
8787 Orion Place
Columbus, OH 43240

ISBN: 978-1-26-421066-4
MHID: 1-26-421066-3

Printed in the United States of America.

1 2 3 4 5 6 7 8 9 LHN 25 24 23 22 21 20

Grade K
Table of Contents

Unit 5

2-Dimensional Shapes

Lessons

Unit 6

Understand Addition

Lessons

Unit 7

Understand Subtraction

Lessons

Unit 8

Addition and Subtraction Strategies

Lessons

Unit 9

Numbers 11 to 15

Lessons

Unit 10

Numbers 16 to 19

Lessons

Unit 11

3-Dimensional Shapes

Lessons

Unit **12**

Count to 100

Lessons

Unit **13**

Analyze, Compare, and Compose Shapes

Lessons

Unit **14**

Compare Measurable Attributes

Lessons

Lesson 2-1 • Reinforce Understanding

Count 1, 2, and 3

Name _____

Review

①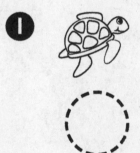

②

Review: There is a counter for each fish. Count. I, 2. There are 2 fish. The counters show 2.

Directions: 1. Count the turtles. Say how many. Trace the counter to show how many. **2.** Count the dolphins. Say how many. Trace the counters to show how many.

Differentiation Resource Book

1

Count 1, 2, and 3

Name _____

Directions: 1. How many shells?. Draw counters to show how many shells. **2.** How many crabs? Draw counters to show how many crabs. **3.** How many snails? Draw counters to show how many snails.

Differentiation Resource Book

2

Represent 1, 2, and 3

Name

Review

2   ③

1   I   2

2   I   2

Review: There is a counter for each rabbit. There are 3 rabbits. The number 3 shows how many rabbits.

Directions: I. Trace a counter for each animal. Say how many. Circle the number to show how many animals. **2.** Trace a counter for each owl. Say how many. Circle the number to show how many owls.

Differentiation Resource Book

Represent 1, 2, and 3

Name _____

1 2

2 1

3 3

Directions: 1–3. Draw counters to show the number.

Count 4 and 5

Name _____

Review

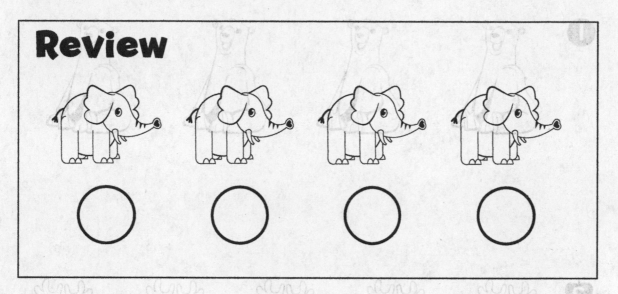

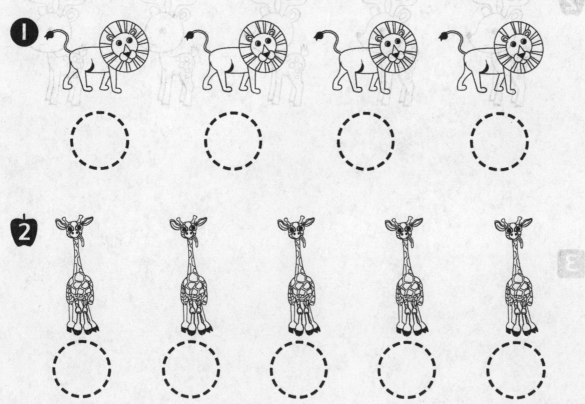

Review: Count the elephants. There are 4 elephants. Four counters show how many elephants.

Directions: 1–2. Count. Say how many. Trace the counters to show how many.

Count 4 and 5

Name _____

Directions: 1–2. Count. Say how many. Draw counters to show how many. **3.** Draw a row of 4 counters. Draw a row of 5 counters. Circle the row with 5 counters.

Represent 4 and 5

Name _____

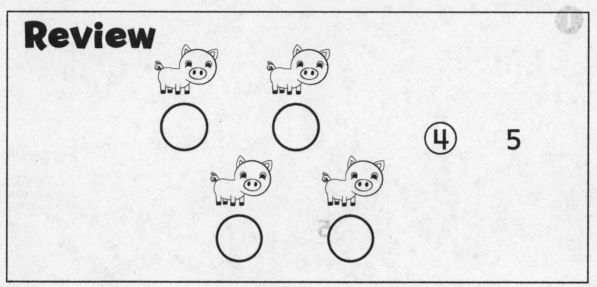

Review

④   5

1   4   5

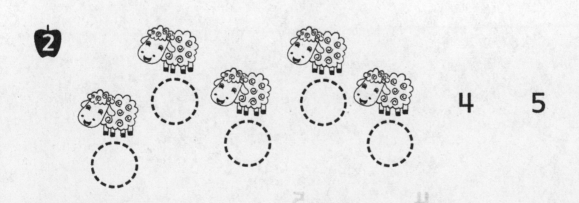

2   4   5

Review: There is a counter for each pig. There are 4 pigs. The number 4 shows how many.

Directions: 1–2. Trace a counter for each animal. Say how many. Circle the number that shows how many.

Represent 4 and 5

Name _____

1

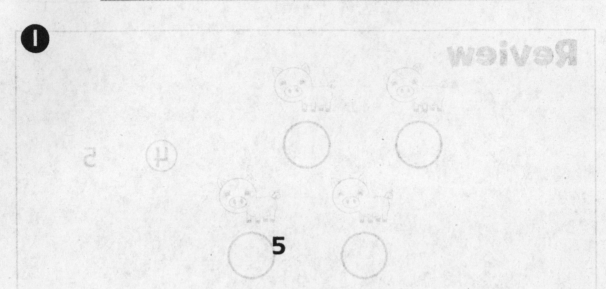

2

4

3

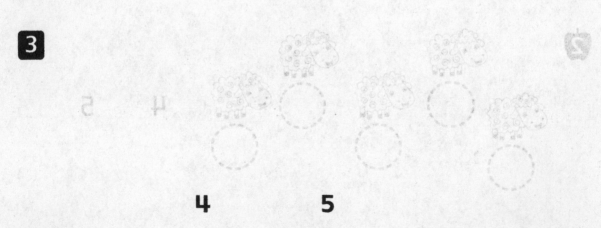

4   5

Directions: 1–2. Draw counters to show the number. **3:** Draw counters to show 4 or 5. Circle the number to show how many.

Differentiation Resource Book

Represent 0

Name _____

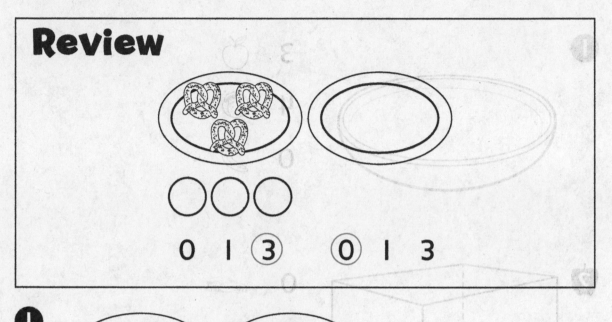

Review

0   1   ③   ⓪   1   3

1

0   1   2   0   1   2

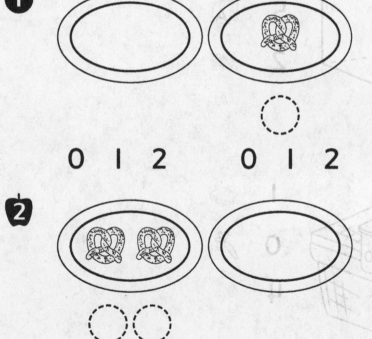

2

0   1   2   0   1   2

Review: Count the pretzels on each plate. There are 3 pretzels on one plate. 3 counters show the pretzels. Three shows the pretzels on the plate. There are 0 pretzels on the other plate. 0 counters show the pretzels. Zero shows the pretzels on the plate.

Directions: 1–2. Count the pretzels on the plate. Draw a counter for each pretzel. Circle the number of pretzels on the plate.

Represent O

Name _____

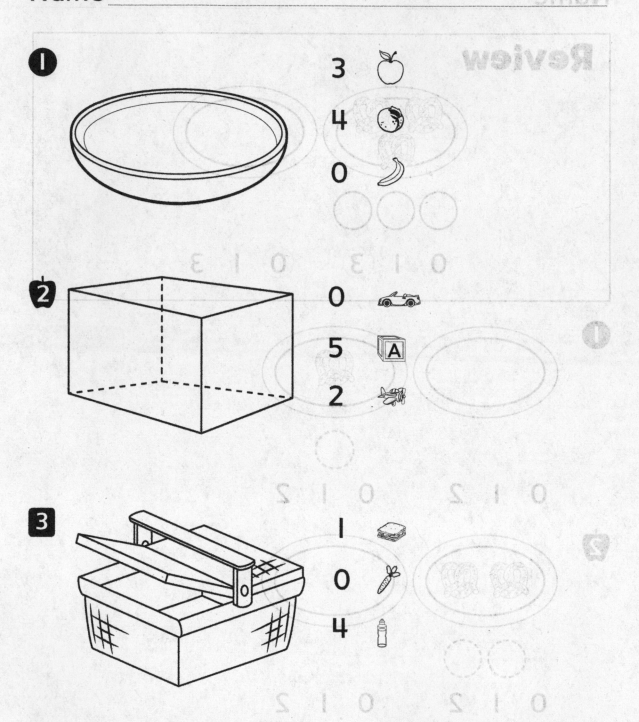

①   3 🍎
         4
         0 🍌

②   0
         5 Ⓐ
         2

③   1
         0
         4

Directions: 1. Draw 3 apples in the bowl. Draw 4 blueberries in the bowl. Draw 0 bananas in the bowl. **2.** Draw 0 cars in the box. Draw 5 blocks in the box. Draw 2 airplanes in the box. **3.** Draw 1 sandwich in the basket. Draw 0 carrots in the basket. Draw 4 bottles in the basket.

Differentiation Resource Book

Numbers to 5

Name _____

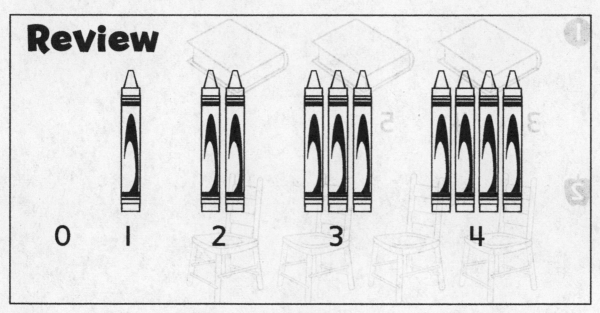

Review

| 0 | 1 | 2 | 3 | 4 |

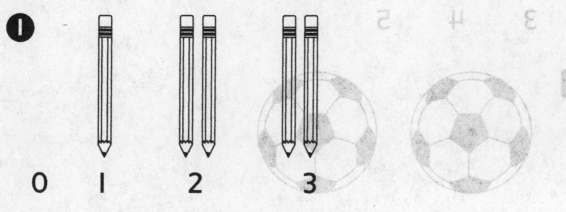

1 0   1   2   3

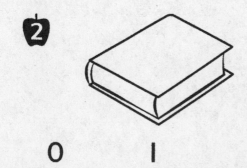

2 0   1   2

Review: One is one more than zero. Two is one more than one. Three is one more than two. Four is one more than three. There are four crayons.

Directions: 1. Draw a pencil to show the number that is one more than two. **2.** Draw books to show the number that is one more than one.

Differentiation Resource Book

II

Numbers to 5

Name _____

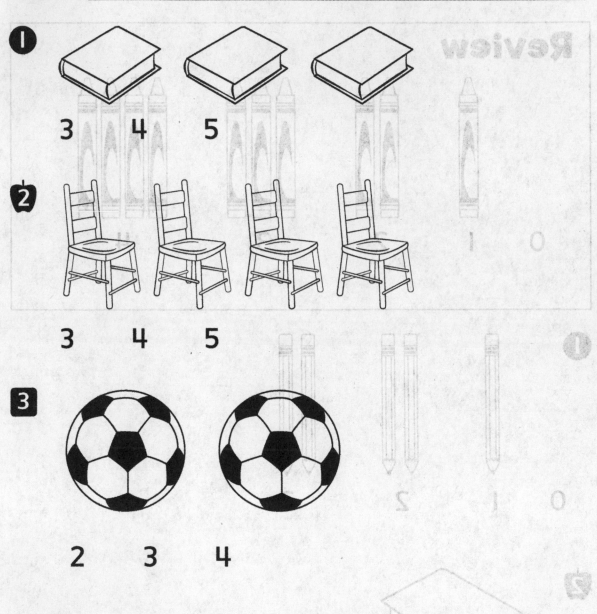

1 3   4   5

2 3   4   5

3 2   3   4

Directions: 1. Count the books. Circle the number that shows 1 more book. **2:** Count the chairs. Circle the number that shows 1 more chair. **3:** Here are Leela's soccer balls. She has 1 more basketball than soccer balls. Circle the number that shows how many basketballs Leela has.

Equal Groups to 5

Name _____

Review

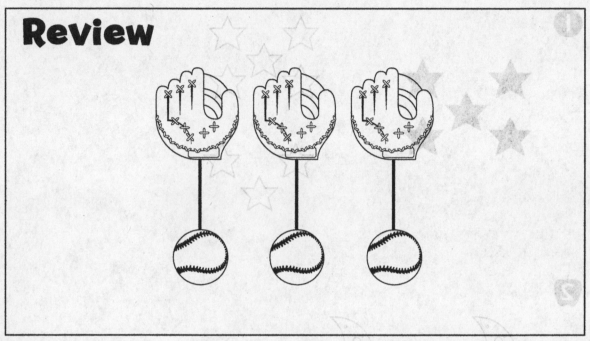

① 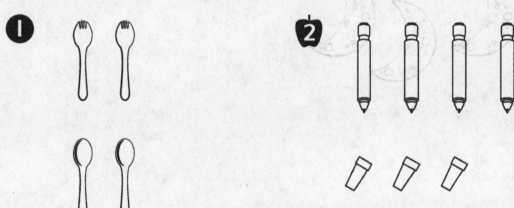 **②**

Review: A line goes from 1 glove to 1 baseball. Each glove matches 1 ball. The groups are equal.

Directions: 1–2. Draw a line from each object to its matching object. Circle the picture that shows equal groups. Draw an X on the picture that does not show equal groups.

Differentiation Resource Book

13

Equal Groups to 5

Name _____

1

2

3

Directions: 1. Count the black stars. Circle the group of white stars that are equal to the number of black stars. Draw an X on the group of white stars that is not equal to the number of black stars. **2.** Count the moons. Draw a group of moons that is not equal. **3.** Vinh draws 5 suns. Draw your own group of suns. Circle them if they are equal to Vinh's suns. Draw an X on them if they are not equal.

Greater Than and Less Than

Name _____

Review

❶

❷

Review: A line goes from 1 pair of glasses to 1 smiley face. The groups are not equal. The group with more is circled. The group with less has an X.

Directions: 1–2. Draw a line from an object in one row to an object in the other row. Circle the group that shows more. Draw an X on the group that shows less.

Differentiation Resource Book

Lesson 2-8 • Extend Thinking

Greater Than and Less Than

Name _____

1

2

3

Directions: 1. Count the fish. Draw a group of fish that is less in the other fishbowl. **2.** Draw 2 flowers. Then draw a group of flowers that has more than 2 flowers. **3.** Rae has 3 leaves. Draw Rae's leaves. Then draw your own group of leaves. Circle your group if it is greater than Rae's group. Put an X on your group if it is less than Rae's group.

Differentiation Resource Book

16

Copyright © McGraw-Hill Education

Compare Numbers to 5

Name _____

Review

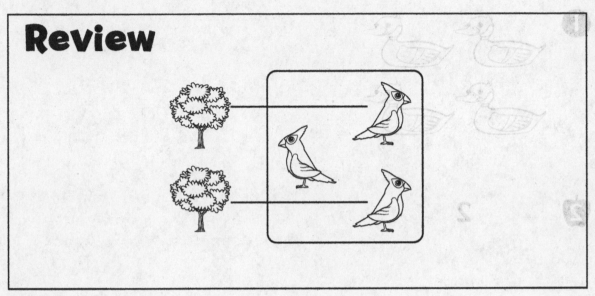

①

②

Review: A line matches I bird with I tree. There are more birds than trees. The group of birds is circled.

Directions: I–2. Draw a line from an object in one group to an object in the other group. Circle the group that is greater.

Lesson 2-9 • Extend Thinking

Compare Numbers to 5

Name

1

2 2

3

Directions: 1. Count the ducks. Draw a group that has fewer ducks. **2.** Draw 2 flowers. Then draw a group that has more flowers. **3.** Count the owls. A group of butterflies is less than the group of owls. Draw the group of butterflies. A group of trees is more than the group of owls. Draw the group of trees.

Differentiation Resource Book

18

Copyright © McGraw-Hill Education

Count 6 and 7

Name _____

Review

❶

❷

❷

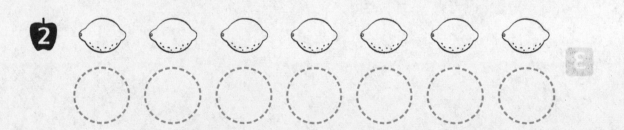

Review: There is 1 counter for each watermelon slice. You can count 1, 2, 3, 4, 5, 6, 7. There are 7 counters. There are 7 watermelon slices.

Directions: 1–2. Count. Say how many. Trace the counters to show how many.

Lesson 3-1 • Extend Thinking

Count 6 and 7

Name _____

①

②

③

Directions: 1. Count the strawberries in each row. Color the row with 7 strawberries.
2. Draw a row with 6 apples. **3.** Draw a row with 7 oranges.

Differentiation Resource Book

Represent 6 and 7

Name _____

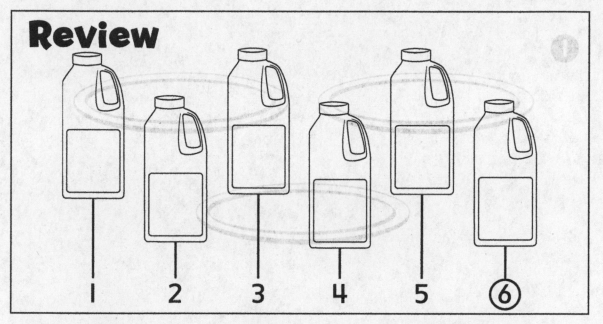

Review

1   2   3   4   5   ⑥

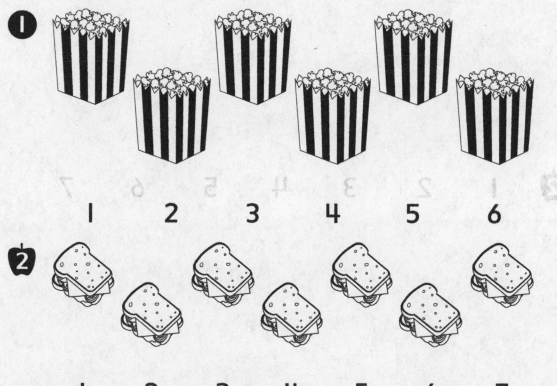

❶

1   2   3   4   5   6

❷

1   2   3   4   5   6   7

Review: You can match numbers to objects as you count. Count the jugs: 1, 2, 3, 4, 5, 6. There are 6 jugs.

Directions: 1–2. Count the objects. Draw a line to match the object with the number as you count. Circle the number that shows how many.

Differentiation Resource Book

Lesson 3-2 • Extend Thinking

Represent 6 and 7

Name _____

1

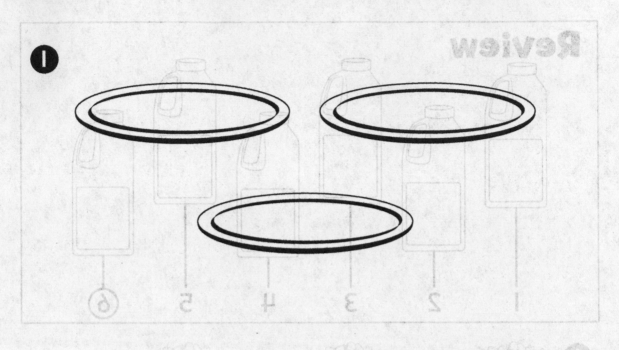

2 1 2 3 4 5 6 7

Directions: Mari buys some fruit. 1. Mari buys 6 apples. Draw the apples on a plate. She buys 7 lemons. Draw the lemons on a plate. She buys 5 strawberries. Draw the strawberries on a plate. **2.** Mari buys 6 or 7 oranges. Draw 6 or 7 oranges. Circle the number of oranges.

Differentiation Resource Book
22

Lesson 3-3 • Reinforce Understanding

Count 8 and 9

Name _____

┌───┐
│ **Review**   ① │
│ 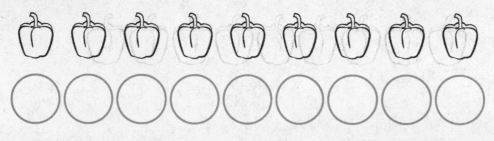   │
└───┘

❶

❷

Review: There is I counter for each pepper. You can count: I, 2, 3, 4, 5, 6, 7, 8, 9. There are 9 counters. There are 9 peppers.

Directions: I–2. Count. Say how many. Trace the counters to show how many.

Differentiation Resource Book

Count 8 and 9

Name _____

①

②

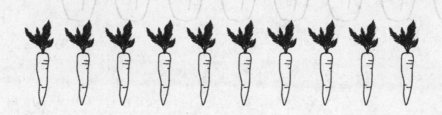

③

④

Directions: Jack buys vegetables. Listen to find out how many of each vegetable he buys. **1.** Jack buys 8 peppers. Circle the peppers Jack buys. **2.** Jack buys 9 carrots. Circle the carrots Jack buys. **3.** Jack buys 5 onion slices. Circle the onion slices Jack buys. **4.** Jack buys 6 ears of corn. Circle the ears of corn Jack buys.

Differentiation Resource Book

Lesson 3-4 • Reinforce Understanding
Represent 8 and 9

Name _____

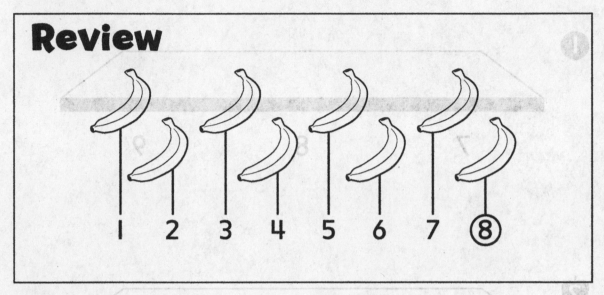

Review

1 2 3 4 5 6 7 ⑧

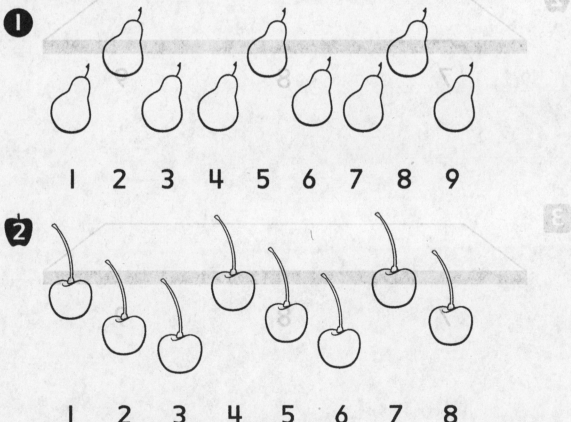

❶

1 2 3 4 5 6 7 8 9

❷

1 2 3 4 5 6 7 8

Review: You can match numbers to objects as you count. Count the bananas: 1, 2, 3, 4, 5, 6, 7, 8. There are 8 bananas.

Directions: 1–2. Count the fruits. Draw a line to match the fruit with a number as you count. Circle the number that shows how many.

Lesson 3-4 • Extend Thinking
Represent 8 and 9
Name _____

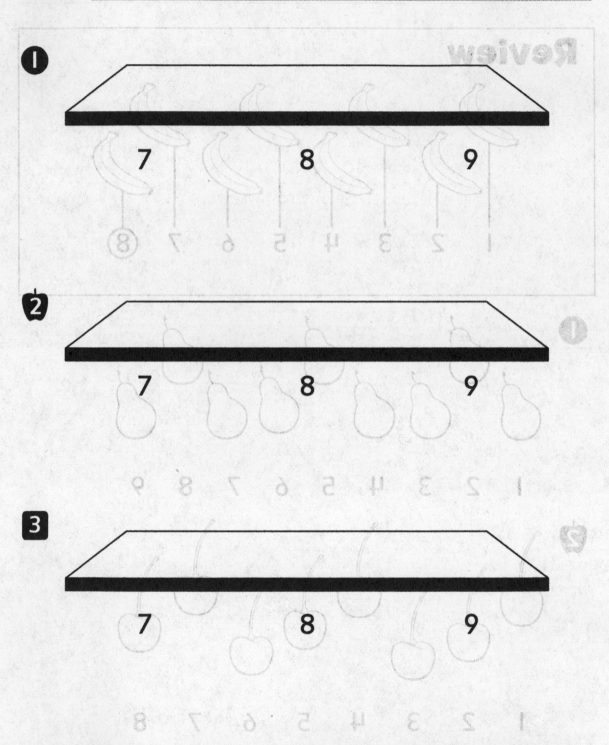

Directions: Jing puts vegetables on shelves at the store. 1. Jing puts 7 carrots on a shelf.
Draw the carrots. Circle the number of carrots. **2.** Jing puts 9 tomatoes on a shelf.
Draw the tomatoes. Circle the number of tomatoes. **3.** Jing put 8 peppers on a shelf.
Draw the peppers. Circle the number of peppers.

Lesson 3-5 • Reinforce Understanding

Count 10

Name _____

Review

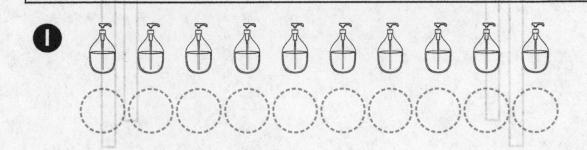

Review: There is 1 counter for each cup. You can count: 1, 2, 3, 4, 5, 6, 7, 8, 9, 10. There are 10 counters. There are 10 cups.

Directions: 1–2. Count. Say how many. Trace the counters to show how many.

Differentiation Resource Book

27

Count 10

Name _____

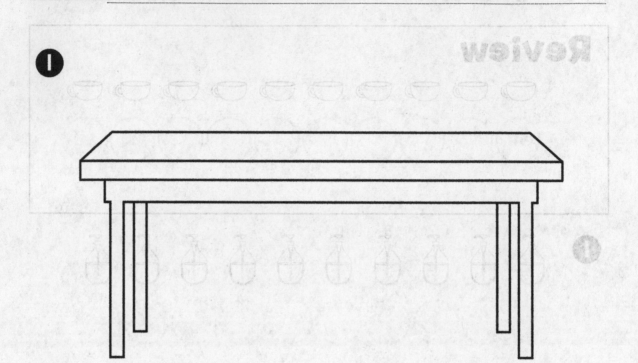

①

②

③

Directions: Finn is getting ready for his soccer game. 1. Finn puts 10 water bottles on the table. Draw the water bottles. **2.** Finn puts 10 soccer balls on the field. Draw the soccer balls. **3.** Finn puts 10 bananas on the table. Draw the bananas.

Lesson 3-6 • Reinforce Understanding

Represent 10

Name

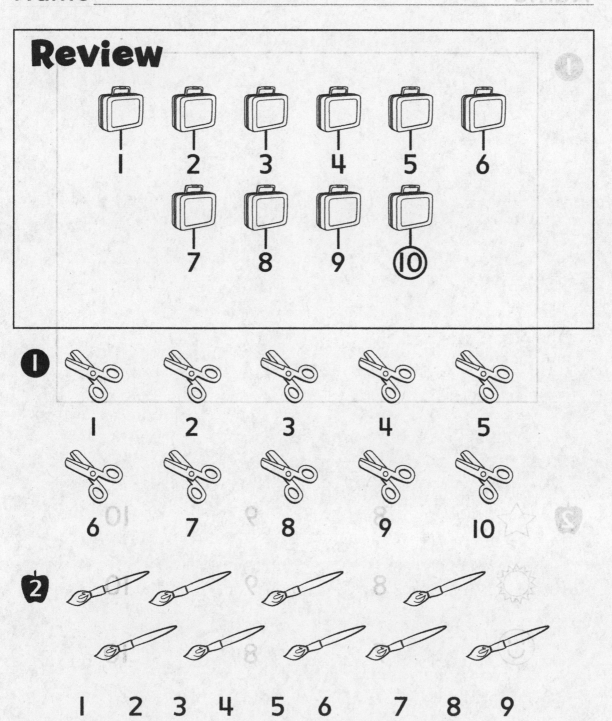

Review: You can match numbers to objects as you count. The last number you count is circled. There are 10 lunch boxes.

Directions: 1–2. Draw a line to match the object with a number as you count. Circle the last number you count.

Represent 10

Name _____

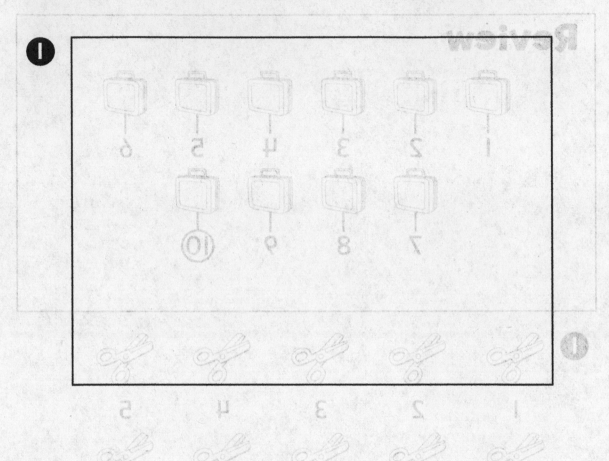

Directions: 1. Alba is drawing a picture. Alba draws 10 stars. Draw the stars. She draws 9 suns. Draw the suns. She draws 7 smiley faces. Draw the smiley faces. **2.** Circle the numbers that tell how many stars, suns, and smiley faces you drew.

Numbers to 10

Name _____

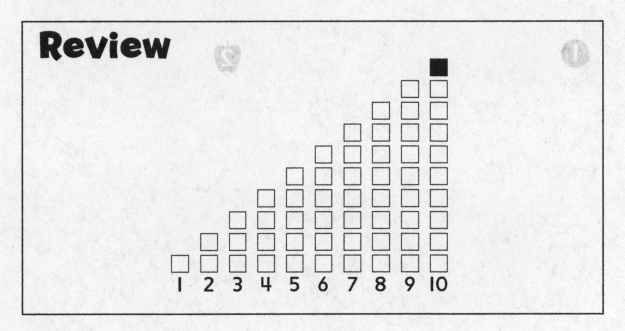

Review

1 2 3 4 5 6 7 8 9 10

❶

5 6 7

❷

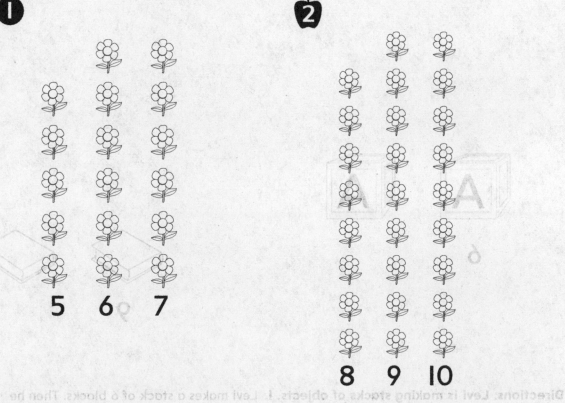

8 9 10

Review: Look at the shapes. Each number shows one more. 10 is one more than 9. It has one more shape.

Directions: 1. Count. Draw one more to show 7. **2.** Count. Draw one more to show 10.

Differentiation Resource Book

Lesson 3-7 • Extend Thinking

Numbers to 10

Name _____

❶

❷

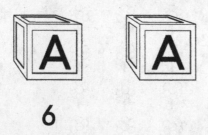

6

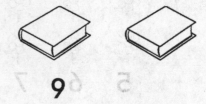

9

Directions: Levi is making stacks of objects. 1. Levi makes a stack of 6 blocks. Then he makes a stack with 1 more block. Draw more blocks to show Levi's stack of 6. Then draw more blocks to show a stack with 1 more. **2.** Levi makes a stack of 9 books. Then he makes a stack with 1 more book. Draw more books to show Levi's stack of 9. Then draw more books to show a stack with 1 more.

Differentiation Resource Book
32

Compare Objects in Groups

Name _____

Review

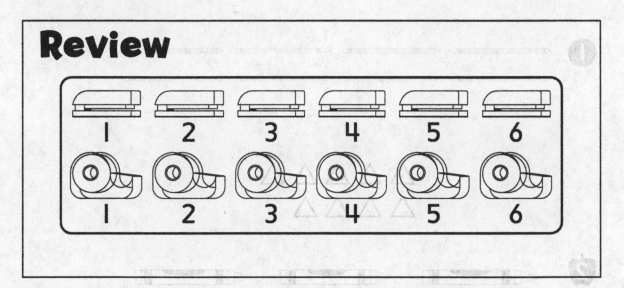

①

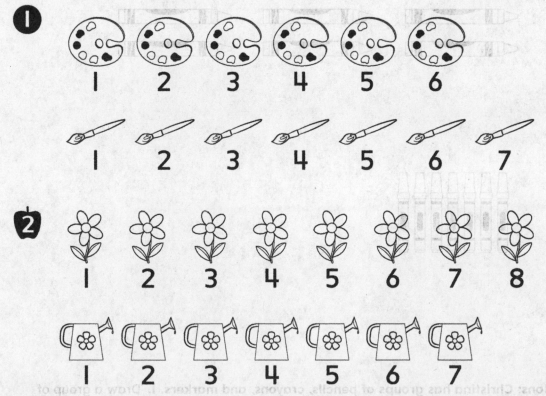

Review: Count the staplers: 1, 2, 3, 4, 5, 6. There are 6 staplers. Count the rolls of tape: 1, 2, 3, 4, 5, 6. There are 6 rolls of tape. Both groups have 6 objects, so the groups are equal.

Directions: 1–2. Count the objects. Circle the group that is greater. Draw an X on the group with less.

Lesson 3-8 • Extend Thinking

Compare Objects in Groups

Name _____

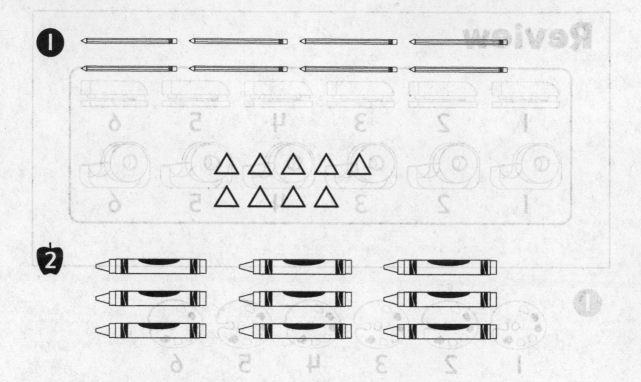

1

2

3

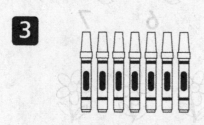

Directions: Christina has groups of pencils, crayons, and markers. 1. Draw a group of objects that is greater than the pencils. Circle the group that is greater. Draw an X on the group that is less. **2.** Draw a group of objects that is less than the crayons. Circle the group that is greater. Draw an X on the group with less. **3.** Draw a group of objects that is equal to the markers. Circle both groups.

Differentiation Resource Book

Compare Numbers

Name

Review

5

⑦

1 8

8

2 9

7

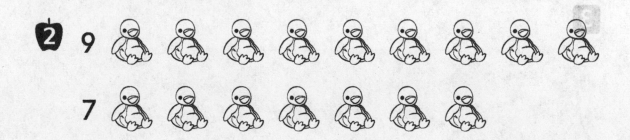

Review: You can draw lines to match the sailboats one-to-one. There are more sailboats in the group of 7 sailboats. The number showing greater group is circled.

Directions: 1–2. Draw lines to match the animals one-to-one. Circle the number in the greater group. If the groups are equal, circle both numbers.

Differentiation Resource Book

Lesson 3-9 • Extend Thinking

Compare Numbers

Name _____

1

1 2 3 4 5 6 7 8 9 10

2

1 2 3 4 5 6 7 8 9 10

3

1 2 3 4 5 6 7 8 9 10

Directions: A farmer looks at his animals. 1. The farmer sees 6 pigs in a pen. He sees more pigs in the barnyard. Draw a number of pigs that he could see in the barnyard. Circle the number of pigs you drew. **2.** The farmer sees 8 horses in the field. He sees fewer horses near the barn. Draw a number of horses that he could see near the barn. Circle the number of horses you drew. **3.** The farmer sees 9 cows in a field. He sees an equal group of cows in the barnyard. Draw the number of cows in the barnyard. Circle the number of cows you drew.

Write Numbers to 3

Name _____

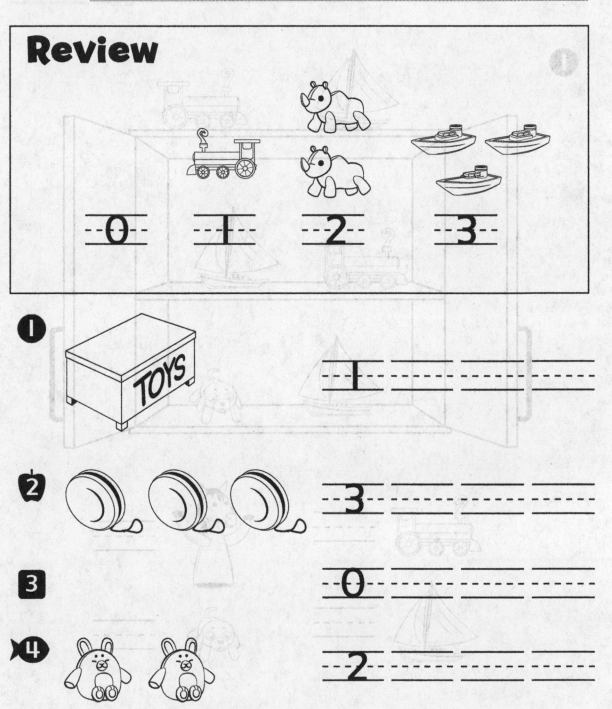

Review

0   1   2   3

1   TOYS   1

2   3

3   0

4   2

Review: You can write numbers to show how many toys. There are 0 dolls. There is 1 toy train. There are 2 toy rhinos. There are 3 toy boats.

Directions: 1–4. Count the objects. Write the number of objects 3 times.

Write Numbers to 3

Name _____

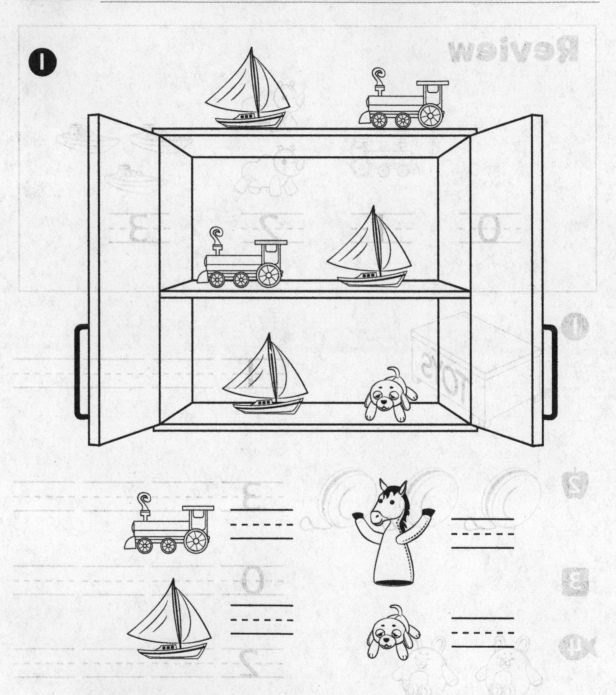

1

Directions: 1. Spencer counts toys at the toy store. Count the number of each toy on the shelves. Write the number of each toy.

Differentiation Resource Book

38

Write Numbers to 6

Name _____

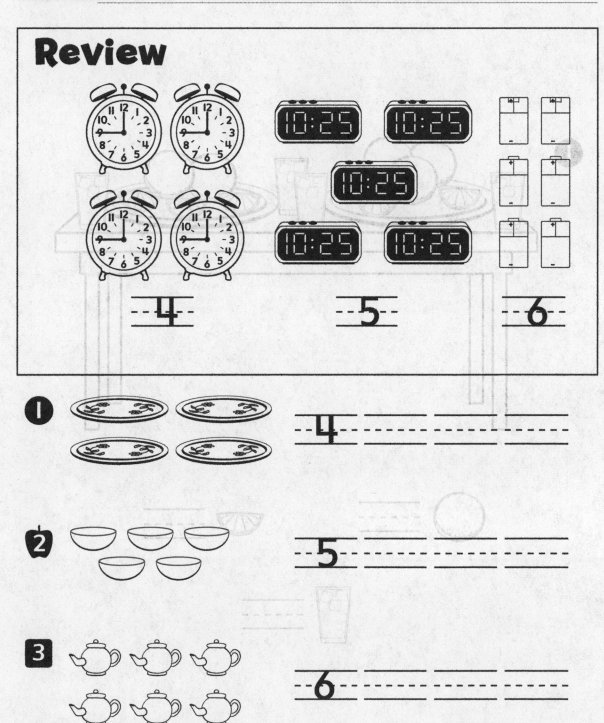

Review You can write numbers to show how many. There are 4 alarm clocks. There are 5 digital clocks. There are 6 batteries.

Directions: 1–3. Count the objects. Write the number of objects 3 times.

Write Numbers to 6

Name

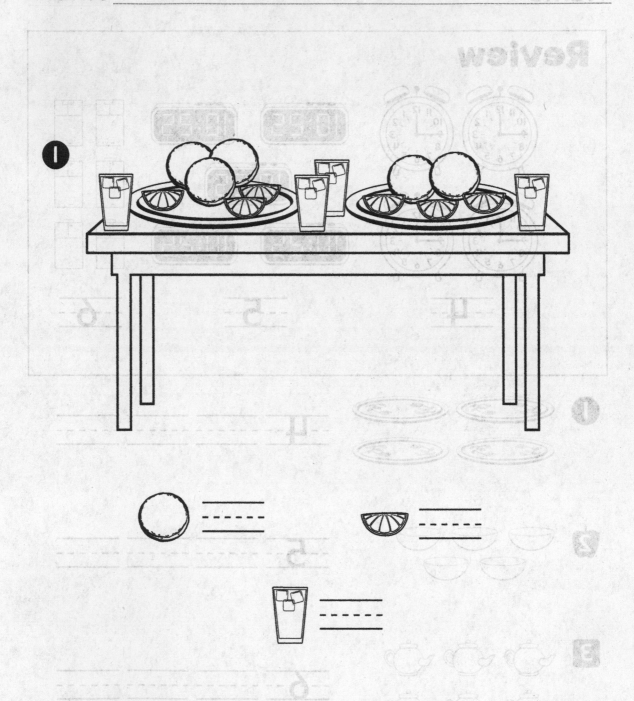

1

Directions: 1. Regina is making breakfast. She puts food on the table. Count the number of each type of food or drink. Write the number of each type of food or drink.

Write Numbers to 10

Name

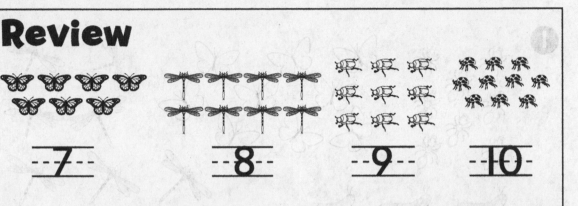

Review

```
--7--        --8--        --9--        --10--
```

1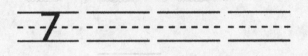

```
--7-- ----- ----- -----
```

2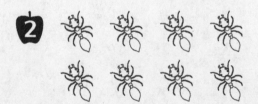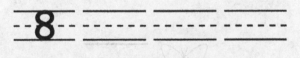

```
--8-- ----- ----- -----
```

3 

```
--9-- ----- ----- -----
```

4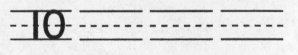

```
--10-- ----- ----- -----
```

Review: You can write numbers to show how many. There are 7 butterflies. There are 8 dragonflies. There are 9 beetles. There are 10 ladybugs.

Directions: 1–4. Count the insects. Write the number of insects 3 times.

Differentiation Resource Book

Lesson 3-12 • Extend Thinking
Write Numbers to 10

Name _____

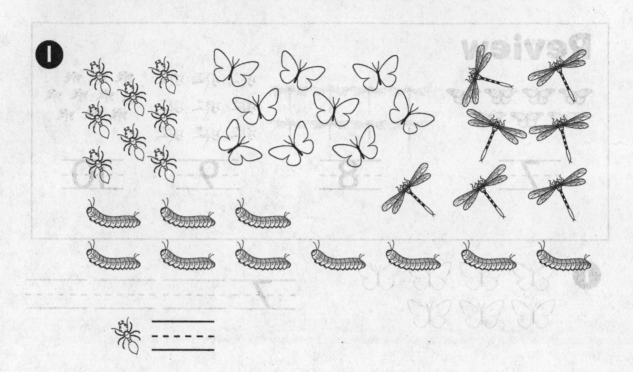

1

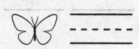
___ ___ ___

___ ___ ___

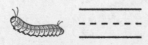
___ ___ ___

___ ___ ___

Directions: 1. Simon is looking at insects. Count the number of each type of insect. Write the number of each type of insect he sees.

Differentiation Resource Book
42

Alike and Different

Name _____

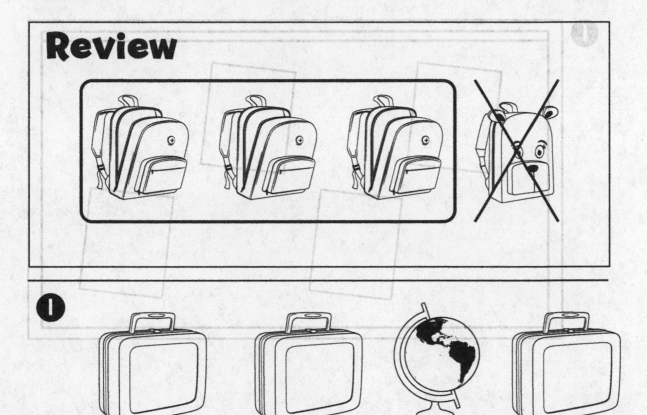

Review

  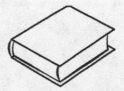 

Review: Objects can be alike or different. One backpack is different. It has a bear on it. An X is drawn on the backpack that is different. The other 3 backpacks are the same. The backpacks that are the same are circled.

Directions: 1-2. Which object is different? Draw an X on the object that is different. Circle the objects that are the same. Explain your thinking.

Differentiation Resource Book

Lesson **4-1** • Extend Thinking

Alike and Different

Name _____

1

Directions: 1. Ms. Chen hangs 4 drawings on a board in the hall. Some of the art is alike. Some of the art is different. Draw the 4 drawings. Which art is alike? Circle the art that is alike. Draw an X on the art that is different. Explain your thinking.

Differentiation Resource Book
44

Sort Objects into Groups

Name _____

Review

1

2

Review: You can sort items into groups by color. All the peppers are shaded. The vegetables that are not shaded are not peppers and have an X drawn on them.

Directions: 1. How can you sort the fruits? Color all the cherries red. Draw an X on all the fruit that is not colored red. **2.** How can you sort the items? Color all the flowers yellow. Draw an X on each item you did not color yellow.

Sort Objects into Groups

Name

Directions: I. Jae sees barnyard animals. Some of the animals are pigs and some are cows. How can you sort the animals by color, size, and shape? Color the animals that are the same size pink. Circle the animal pairs that are the same shape. Draw an X on the animals that are the same color. Draw another animal that could be in the same group with the pigs or cows. Draw a line from your animal to the group it belongs to. Explain your thinking.

Differentiation Resource Book
46

Count Objects in Groups

Name _____

Review

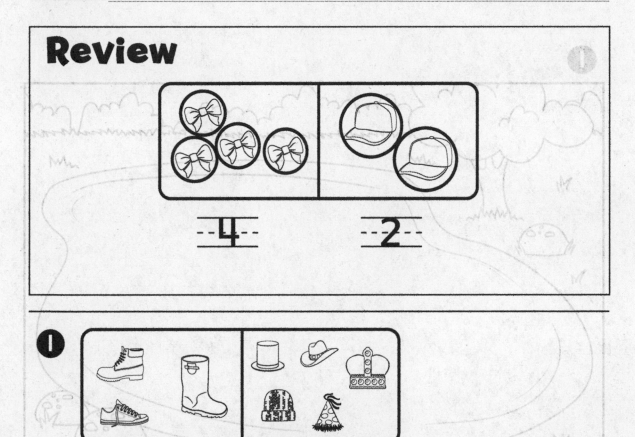

----4----   ----2----

❶
_ _ _ _ _ _   _ _ _ _ _ _

❷

_ _ _ _ _ _   _ _ _ _ _ _

Review: You can count objects in a group. The bows are circled. Count each bow: 1, 2, 3, 4. Four bows are the group. The hats are circled. Count each hat: 1, 2. Two hats are the group.

Directions: How can you count objects in a group? 1. Circle each shoe as you count. Write a number to show how many. Circle each hat as you count. Write a number to show how many. **2.** Circle each pair of glasses as you count. Write a number to show how many. Circle each sweater as you count. Write a number to show how many.

Lesson 4-3 • Extend Thinking
Count Objects in Groups

Name _____

①

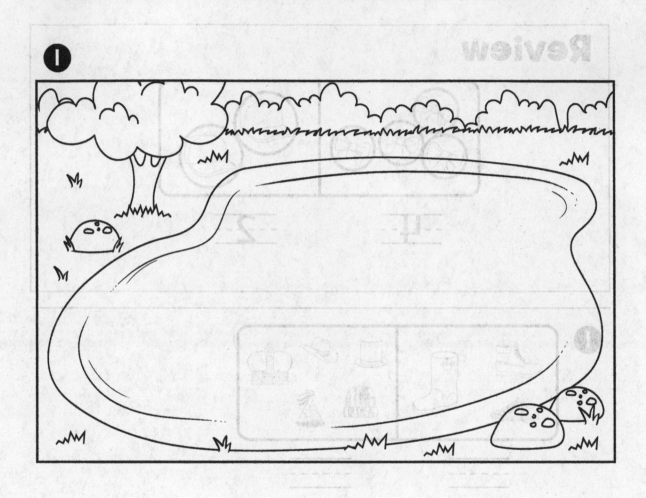

② – – – – – – – –

Directions: Peter is at the pond. He sees 3 groups of fish. **1.** Draw 3 groups of fish. Circle each group. **2.** How can you show how many fish are in each group? Count the number of fish in each group. Write the numbers to show how many. Draw a line from the group of fish to the number that tells how many.

Differentiation Resource Book
48

Describe Groups of Objects

Name _____

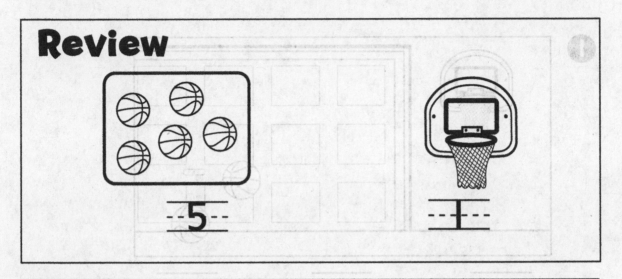

Review

5

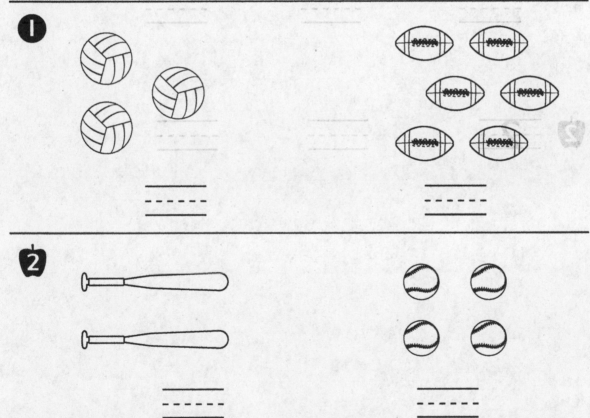

1

_ _ _ _ _ _   _ _ _ _ _ _

2

_ _ _ _ _ _   _ _ _ _ _ _

Review: You can count the number of objects in a group. Count the hoops: I. There is I hoop. Count the basketballs: I, 2, 3, 4, 5. The group of basketballs has more. The group with more is circled.

Directions: How can you count the objects in the group? I. Circle the balls in the group as you count. Write the number in the group. Circle the group with fewer. **2.** Circle the bats or balls in the group as you count. Write the number in the group. Circle the group with more.

Differentiation Resource Book

Describe Groups of Objects

Name _____

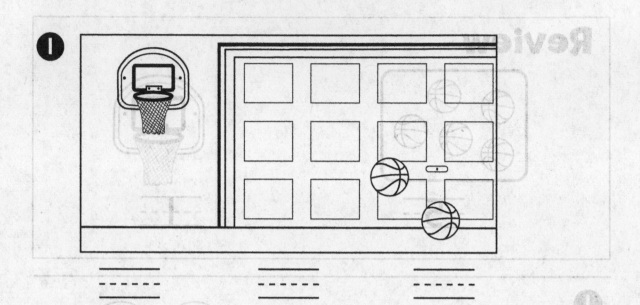

1

2 _ _2_ _

Directions: Megan and Kiri play basketball and soccer with friends after school.
1. Count the hoop in the group. Write the number. Count the group of basketballs. Write the number. Draw a group of more friends than the number of basketballs. Write the number of friends. **2.** Draw a group of soccer balls to match the number. Draw a group of more friends than soccer balls. Write the number. How can you draw a group of water bottles that shows the most? Write the number.

Differentiation Resource Book
50

Lesson 5-1 • Reinforce Understanding

Triangles

Name

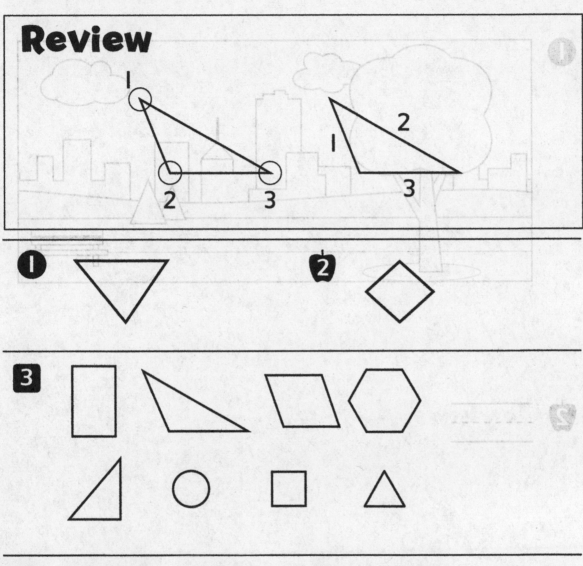

Review

Review: Count the vertices. 1, 2, 3. Count the sides. 1, 2, 3. The shape has 3 vertices and 3 sides. It is a triangle.

Directions: 1–2. How many vertices? Count and circle the vertices. How many sides? Count and trace the sides. Draw an X on the shape if it is not a triangle. **3.** Which shapes are triangles? Circle the shapes that are triangles. Draw an X on the shapes that are not triangles. **4:** Which objects are shaped like a triangle? Circle the objects that are shaped like a triangle. Draw an X on the objects that are not shaped like a triangle.

Triangles

Name _____

1.

2.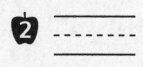

Directions: Miles plays at the park. He sees some triangles. 1. Which objects are shaped like triangles? Circle the objects that are shaped like triangles. Draw two triangles in the picture. Circle your triangles. **2.** How many triangles? Write the number that tells how many triangles are in the picture. Explain your thinking.

Squares and Rectangles

Name

Review

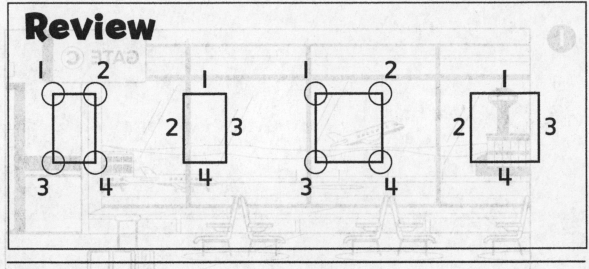

Review: Count the vertices on each shape. l, 2, 3, 4. Count the sides on each shape. l, 2, 3, 4. The shapes have 4 vertices and 4 sides. Both shapes are rectangles. The shape on the right is a kind of rectangle that has 4 equal sides. It is called a square.

Directions: l–2. How many vertices? Count and circle the vertices. How many sides? Count and trace the sides. Color the square purple. Color the rectangle green. **3–4:** Which objects are shaped like rectangles? Color the rectangles your favorite colors. Circle the square.

Differentiation Resource Book

Squares and Rectangles

Name

1

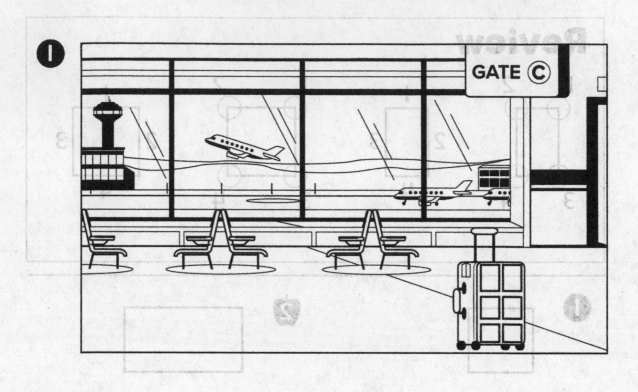

GATE C

2

Directions: Gia is at the airport. She sees some shapes. 1. Which are there more of, objects shaped like triangles, rectangles, or squares? Explain your thinking. **2.** Draw a shape in the picture. Put an X on your shape if it is a rectangle. Circle your shape if it is not a rectangle. Explain your thinking.

Lesson 5-3 • Reinforce Understanding

Hexagons

Name _____

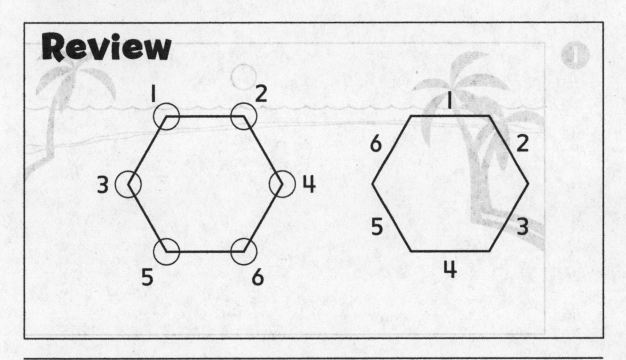

Review

Review: Count the vertices. 1, 2, 3, 4, 5, 6. Count the sides. 1, 2, 3, 4, 5, 6. The shape has 6 vertices and 6 sides. It is a hexagon.

Directions: 1–2. How many vertices? Count and circle the vertices. Circle the shape if it is a hexagon. Draw an X on the shape if it is not. **3:** Circle the hexagons. Draw an X on the shape if is it is not a hexagon.

Hexagons

Name _____

Directions: Luis is at the beach. He draws 4 shapes in the sand. He draws rectangles, triangles, and hexagons. **1.** Which shapes could Luis have drawn? Draw the shapes. Color any rectangles green. Color any triangles orange. Color any hexagons blue. **2.** Which shape did you draw the most? Name and circle the shape you drew the most.

Lesson 5-4 • Reinforce Understanding

Circles

Name _____

Circles

Name

Review

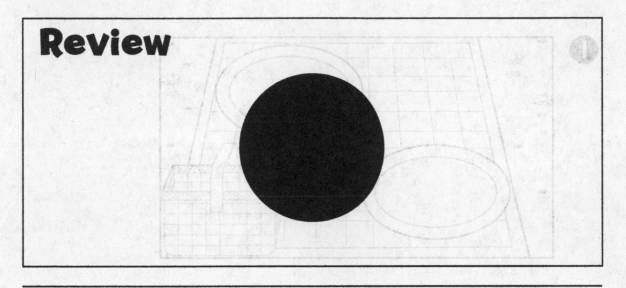

①

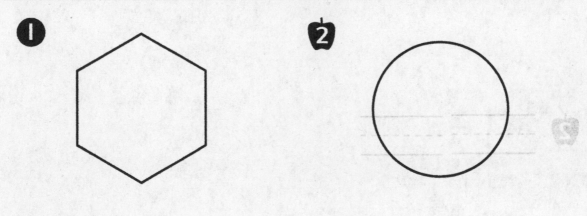

3

Review: This shape has 0 vertices. This shape has 0 sides. It is a circle.

Directions: 1–2. How many vertices? Count and circle the vertices. Circle the shape if it is a circle. Draw an X on the shape if it is not. **3:** Circle the circles. Draw an X on the shape if is it is not. **4:** Circle the objects shaped like circles. Draw an X on the shape if is it is not.

Circles

Name

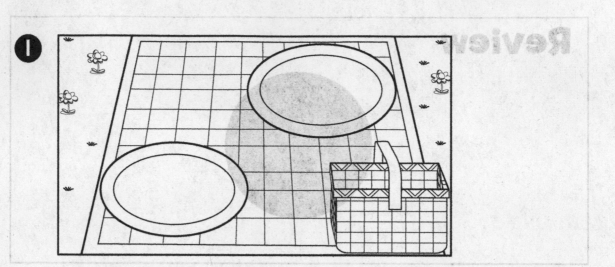

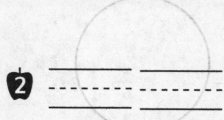

Directions: Anya is getting ready for a picnic. Anya puts foods shaped like a circle on one plate. She puts foods that are not shaped like a circle on the other plate. **1.** Draw some food shaped like a circle on one plate. Draw some food not shaped like a circle on the other plate. **2.** How many foods are shaped like a circle? Count the number of foods shaped like a circle. Write the number. Circle it to show that it tells the number of foods shaped like a circle. Count the number of foods not shaped like a circle. Write the number. Describe the shapes.

Position of 2-Dimensional Shapes

Name _____

Review

1

2

Review: The circle is *above* the triangle. The rectangle is *below* the triangle. The square is *next* to the triangle. The hexagon is *behind* the triangle.

Directions: 1: Which trees are in *front*? Color the trees green that are in *front*. Which trees are *behind*? Color the trees blue that are *behind*. **2:** How can you draw a shape *above* the animals? Explain your thinking.

Differentiation Resource Book

Position of 2-Dimensional Shapes

Name

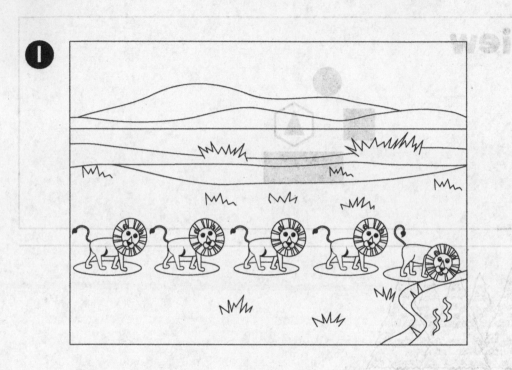

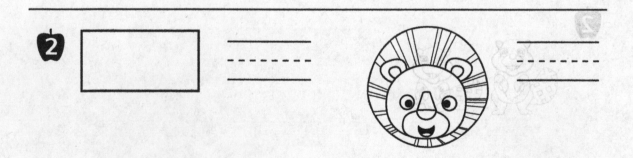

Directions: Lions are at the watering hole. **1.** Draw a square on the lion that is *in front* of all the other lions. Draw a triangle on the lion that is *behind* all the other lions. Draw a sun in the sky *above* all the lions. Draw more grass *below* the lions. Draw rectangles *next* to the water. How did you know where to draw the shapes? **2.** Count the number of rectangles you drew. Write the number. Count the number of lions. Write the number. Circle the number that is more.

Differentiation Resource Book

Represent and Solve Add To Problems

Name _____

Review

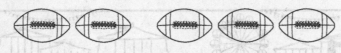

2 and 3 is 5.

1

3   and   1   is _____.

2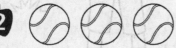

3   and   3   is _____.

3

1   and   4   is _____.

Review: Hannah has two footballs. She gets three more footballs. How many footballs does Hannah have now? Count all the footballs to find how many. 1, 2, 3, 4, 5. There are five footballs. Two and three is five.

Directions 1–3: How many? Count the number in one group. Count the number in the other group. Write the total.

Lesson 6-1 • Extend Thinking

Represent and Solve Add To Problems

Name _____

1

2 _____ _____

_____ and _____ is _____.

Directions: Grayson and Maria have seven balloons. Create a word problem to show how many balloons Grayson and Maria have. 1. How many balloons do the children have? Draw Grayson's and Maria's balloons. **2.** How many balloons do they have? Count Grayson's balloons. Write the number. Count Maria's balloons. Write the number. Add Maria's balloons to Grayson's balloons. Write the number that tells how many balloons in all.

Differentiation Resource Book
62

Represent and Solve More Add To Problems

Name _____

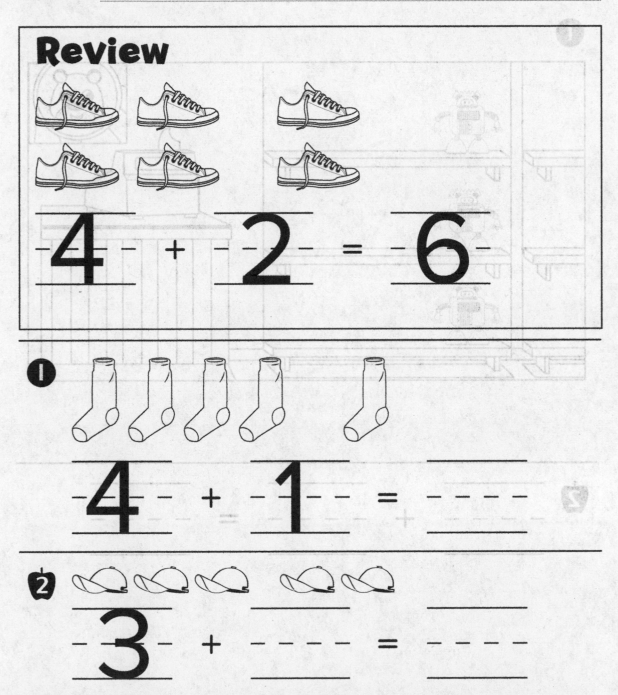

Review

$$4 + 2 = 6$$

❶ $$4 + 1 = \underline{}$$

❷ $$3 + \underline{} = \underline{}$$

Review: You can use drawings to represent an addition story. Then you can write an equation. Nick sees four shoes. He sees two more shoes. How many shoes does Nick see? Count all the shoes to find how many. 1, 2, 3, 4, 5, 6. There are six shoes. Four plus two equals six.

Directions 1–2: How many? Count the number in each group. Write how many. How many in both groups? Write the number.

Represent and Solve More Add To Problems

Name

1

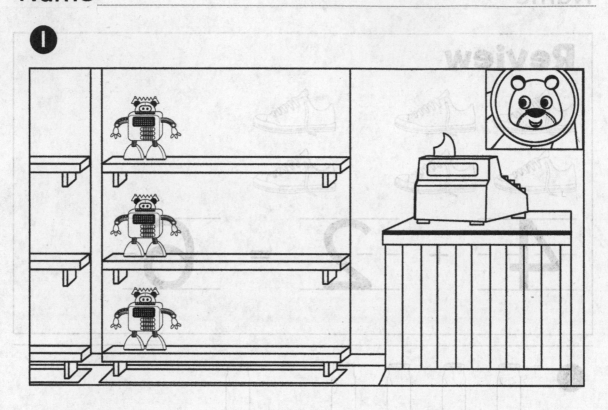

2 _____ + _____ = _____

Directions: Kyo visits a toy store. He sees some robots. 1. Draw one to six more robots on the shelves. **2.** Count the robots shown on the shelves. Write the number. Count the robots you drew. Write the number. How many robots now? Write the number.

Differentiation Resource Book
64

Lesson 6-3 • Reinforce Understanding
Represent and Solve Put Together Problems

Name _____

Review

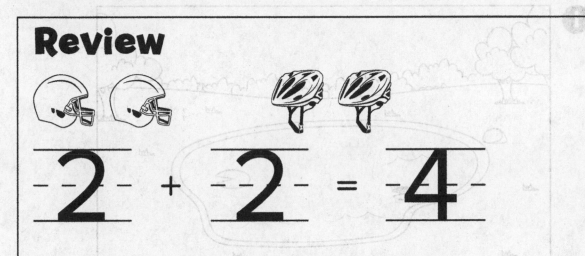

$$2 + 2 = 4$$

1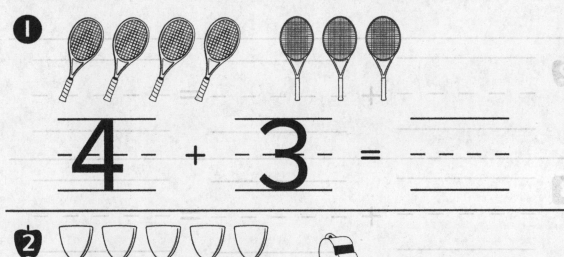

$$4 + 3 = \text{-----}$$

2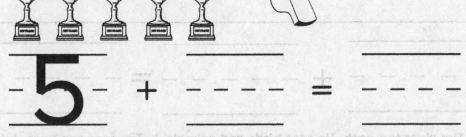

$$5 + \text{-----} = \text{-----}$$

Review: You can use drawings to represent an addition story. Then you can write an equation. Rika sees two football helmets. She sees two bicycle helmets. How many helmets does Rika see? Count all the helmets to find how many. 1, 2, 3, 4. There are four helmets. Two plus two equals four.

Directions 1–2: Count the number in each group. Write how many. How many in both groups? Write the number.

Represent and Solve Put Together Problems

Name

1

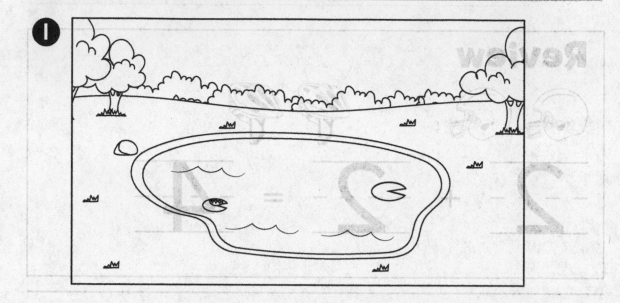

2

$$\underline{\hspace{4cm}} + \underline{\hspace{4cm}} = \underline{\hspace{4cm}}$$

3

$$\underline{\hspace{4cm}} + \underline{\hspace{4cm}} = \underline{\hspace{4cm}}$$

4

$$\underline{\hspace{4cm}} + \underline{\hspace{4cm}} = \underline{\hspace{4cm}}$$

Directions: Tim visits the park. He sees birds and animals. 1. Tim sees one swan, two snails, and three ladybugs. Draw the animals Tim sees. **2.** How many swans and snails does Tim see? How can you write an equation to show the swan and snails? **3.** How many snails and ladybugs does Tim see? How can you write an equation to show the snails and ladybugs? **4.** How many swans and ladybugs does Tim see? How can you write an equation to show the swan and ladybugs?

Differentiation Resource Book

Lesson 6-4 • Reinforce Understanding
Represent and Solve Addition Problems

Name _____

Lesson 6-4 • Exten

Review

 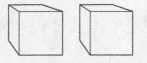

$3 + 2 = 5$

$2 + 3 = 5$

---

①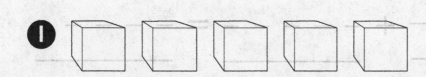

_____   _____

- - - - - + - - - - - = - - - - -

_____   _____

---

②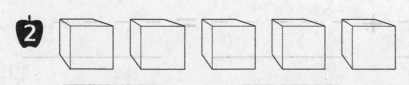

_____   _____

- - - - - + - - - - - = - - - - -

_____   _____

Review: Erik can make rows of five blocks. First, he uses three letter blocks and two plain blocks. Then he uses two letter blocks and three plain blocks.

Directions: 1. Color some blocks red. Color the rest of the blocks green. Then write an equation. **2.** Show a different way to make five. Color some blocks red. Color the rest of the blocks green. Then write an equation.

Differentiation Resource Book

Represent and Solve Addition Problems

Name _____

1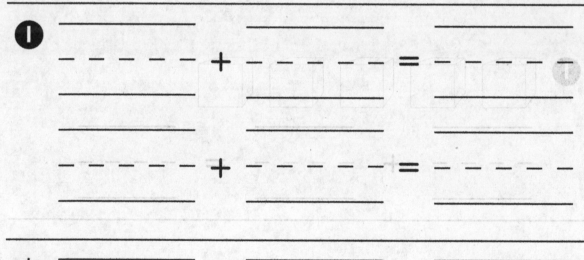

_ _ _ _ _ + _ _ _ _ _ = _ _ _ _ _

_ _ _ _ _ + _ _ _ _ _ = _ _ _ _ _

2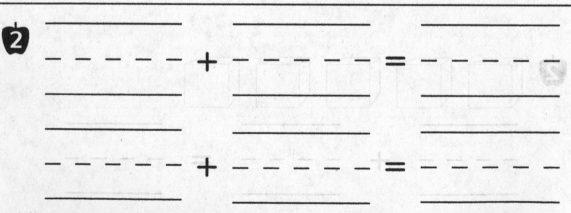

_ _ _ _ _ + _ _ _ _ _ = _ _ _ _ _

_ _ _ _ _ + _ _ _ _ _ = _ _ _ _ _

Directions: Bianca plays with three toy dinosaurs. 1. Draw Bianca's dinosaurs. Write two different related equations to show how many dinosaurs Bianca plays with. **2.** Bianca gets two new dinosaurs. Draw them. Write two different related equations to show how many dinosaurs Bianca has now.

Represent and Solve More Addition Problems

Name _____

Review

$$3 + 3 = 6$$
$$2 + 4 = 6$$
$$1 + 5 = 6$$

❶ _____ + _____ = _____

❷ _____ + _____ = _____

Review: DeMarcus has six trains. He writes three equations about his trains. They show different ways to make six.

Directions 1–2: Write a new equation about the trains. Explain how you know the equation shows six.

Represent and Solve More Addition Problems

Name

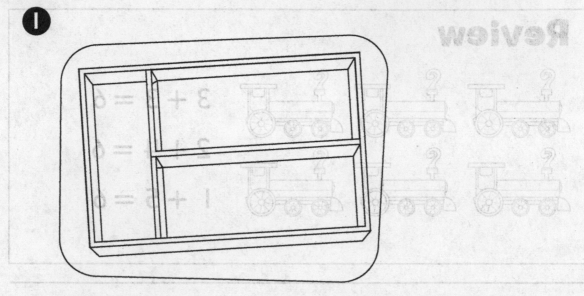

2 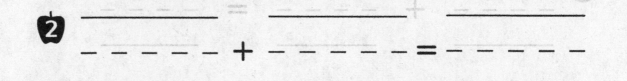 _____ + _____ = _____

3  _____ + _____ = _____

4  _____ + _____ = _____

Directions: You have two or three crayons, three or four pencils, and one eraser in your pencil case. 1. Draw your crayons, pencils, and erasers. **2.** How many crayons and pencils do you have? Write an equation. **3.** How many pencils and erasers do you have? Write an equation **4.** How many erasers and crayons do you have? Write an equation.

Represent Take Apart Problems

Name _____

Review

4   and   2

1

_ 3 _   and   _ _ _

2

_ _ _   _ _ _

and

Review There are 6 keys. One way to break apart 6 is into parts of 4 and 2. There are 4 keys in one part. There are 2 keys in the other part.

Directions: 1. How can you show one way to break apart 6? There are 3 in one part. Circle the number of keys in the other part. Write the number in the other part. **2.** How can you show another way to break apart 6? Circle to show each part. Write the number in each part.

Represent Take Apart Problems

Name _____

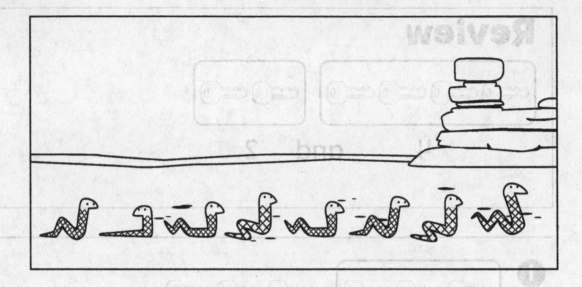

1

_____ _____

_____ **and** _____

_____

2

_____ _____

_____ **and** _____

Directions: Eight snakes crawl on the ground in the desert. 1. How could you take apart the group of snakes to show two parts? Draw the snakes in each part. Write the number of snakes in each part. **2.** All of the snakes crawl in front of the rocks. Draw the snakes in each part now. Write the number of snakes in each part.

Differentiation Resource Book

72

Represent and Solve Take From Problems

Name

Review

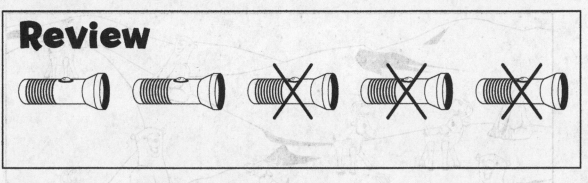

①

②

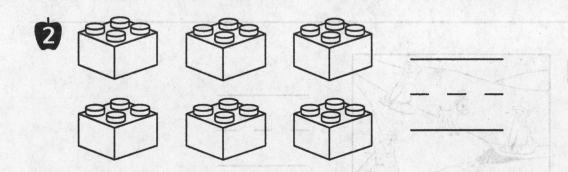

Review: You can use drawings to show a subtraction story. Five flashlights are at a store. Tyler buys 3 flashlights. How many flashlights are left? An X marks a flashlight that was sold. Count the number that are left. There are 2 flashlights left.

Directions: 1. Vince has 8 bears. He gives 4 bears away. How many bears are left? Draw an X on the bears that Vince gives away. Count the bears that are left. Write the number. **2.** Shay has 6 blocks. She gives 3 blocks away. How many blocks are left? Draw an X on the blocks Shay gives away. Count the blocks that are left. Write the number.

Represent and Solve Take From Problems

Name

①  _____

② _____

Directions: Reindeer and polar bears are in the snow. 1. Tell a story where some of the reindeer choose to leave the group. Draw an X on the reindeer that choose to leave the group. Write the number that are left. **2.** Tell a story where some of the polar bears choose to leave the group. Draw an X on the polar bears that leave the group. Write the number that are left. **3.** Tell a story where some more of the animals choose to leave the group. Draw your story. Write the number of animals that are left.

Differentiation Resource Book

74

Represent and Solve More Take From Problems

Name _____

Review

$6 - 2 = 4$

1

$\underline{}7\underline{} - \underline{}4\underline{} = \underline{}$

2

$\underline{}6\underline{} - \underline{} = \underline{}$

Review: James has 6 toy boats. He gives 2 away. How many boats are left? An X marks the boats that James gives away. Count the number that are left. 1, 2, 3, 4. James has 4 toy boats left. The subtraction equation 6 minus 2 equals 4 represents the picture.

Directions: 1. Nina has 7 stuffed penguins. She gives 4 away. How many stuffed penguins are left? Draw an X on the penguins Nina gives away. Write numbers to complete the equation.
2. Raj has 6 stuffed rabbits. He gives 1 away. How many stuffed rabbits are left? Draw an X on the rabbits Raj gives away. Write numbers to complete the equation.

Differentiation Resource Book

75

Represent and Solve More Take From Problems

Name _____

1

_____ _____ _____

_____ — _____ = _____

2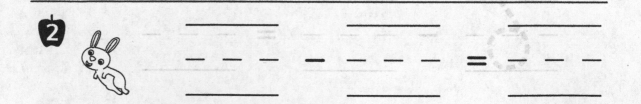

_____ _____ _____

_____ — _____ = _____

Directions: Leo sees birds and rabbits at the park. 1. Tell a story where some of the birds leave the group. Draw an X on the birds that leave the group. Write a subtraction equation to show how many are left. **2.** Tell a story where some of the rabbits leave the group. Draw an X on the rabbits that leave the group. Write a subtraction equation to show how many are left.

Differentiation Resource Book

76

Represent and Solve Subtraction Problems

Name _____

Review

$$8 - 7 = 1$$

1

$$7 - 5 = \text{---}$$

2

$$9 - \text{---} \,\, = \text{---}$$

Review: Mia has 8 orange slices. She gives 7 away. How many orange slices are left? An X marks a slice that Mia gives away. Count the number that are left. Mia has 1 orange slice left. The subtraction equation 8 minus 7 equals 1 represents the picture.

Directions: 1. Milo has 7 pretzels. He gives 5 away. How many pretzels are left? Draw an X on the pretzels Milo gives away. Count the number that are left. Write numbers and trace the symbols to complete the equation. **2.** Izzy has 9 bananas. She gives 4 away. How many bananas are left? Draw an X on the bananas Izzy gives away. Count the number that are left. Write numbers and trace the symbols to complete the equation.

Represent and Solve Subtraction Problems

Name

❶ _____

- -

_____

 2 _____

- -

_____

Directions: Deandre sells peppers at a stand. 1. Tell a story where Deandre sells some peppers to Stella. Draw Xs on the pepper to show what Stella buys. Write a subtraction equation to represent your story. Be sure to include the subtraction sign and equal sign. **2.** Tell a story where Deandre sells some of the peppers that are left to Sean. Circle the peppers to show what Sean buys. Write a subtraction equation to represent your story. Be sure to include the subtraction sign and equal sign.

Lesson **7-5** • Reinforce Understanding

Represent and Solve Addition and Subtraction Problems

Name _____

Review

$$3 + 2 = 5$$

$$5 - 2 = 3$$

1

2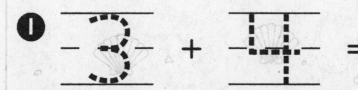

Review: Kim finds 3 sand dollars. Then she finds 2 more. How many sand dollars does Kim have? The addition equation 3 plus 2 equals 5 represents the picture. Count. 1, 2, 3, 4, 5. Kim has 5 sand dollars. She gives 2 to Derek. How many sand dollars does Kim have left? Count. 1, 2, 3. The subtraction equation 5 minus 2 equals 3 represents the picture.

Directions: 1. Cruz finds 3 shells. Then he finds 4 more shells. How many shells does Cruz have? Count the shells. Write numbers and trace the signs to complete the addition equation. **2.** Cruz has 7 shells. He gives 4 away. How many shells does Cruz have left? Count the shells. Write numbers and trace the signs to complete the subtraction equation. Explain how you know when to use addition and when to use subtraction.

Differentiation Resource Book

Represent and Solve Addition and Subtraction Problems

Name _____

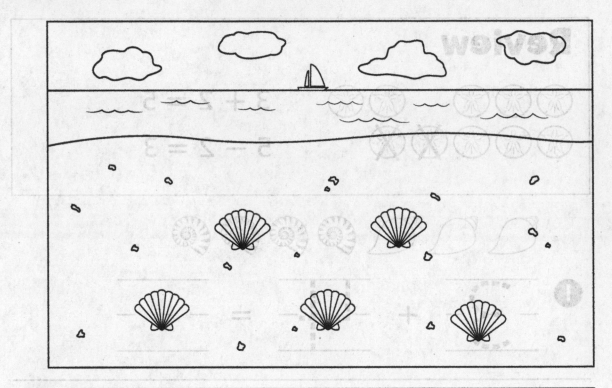

① _____

② _____

Directions: Sarah finds 5 shells at the beach. 1. Sarah finds 4 more shells on the beach. Draw the shells Sarah finds. Write an addition equation to represent your story. Be sure to include the signs. **2.** Sarah gives some shells away. Draw an X on the shells Sarah gives away. Write a subtraction equation to represent your story. Be sure to include the signs.

Add within 5

Name _____

Review

 $2 + 1 = 3$

1

$2 + 2 = \text{---}$
_____

2

_____
$1 + 3 = \text{---}$

Review: You can use counters to add. The addition equation for this is 2 plus 1 equals 3.

Directions: 1-2. How can you use counters to find the sum? Look at the counters in each group. Then write the sum.

Add within 5

Name _____

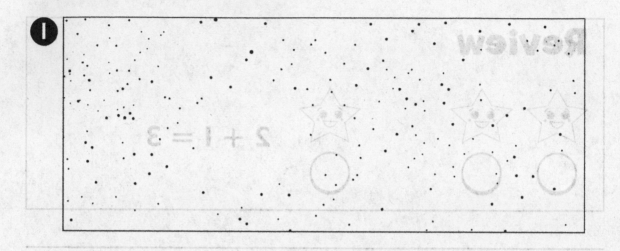

1

$2 + 2 =$ _ _ _ _

⭕ ⭕ ⭕ ⭕

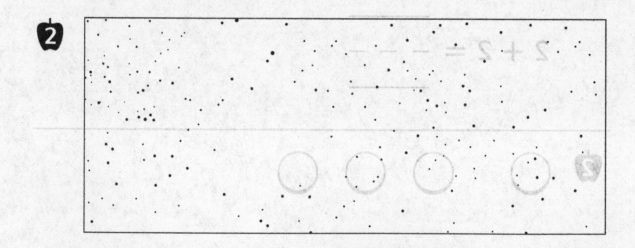

2

$3 + 1 =$ _ _ _

Directions: There are some rockets in space. **1.** Draw groups of rockets to match the equation. Write the total. **2.** Draw groups of rockets to match your equation. Tell a story for your equation. Write the total.

Subtract within 5

Name

Review

4 − 3 = 1

① ○ ○ ○ ○ ○

5 − 2 = _ _ _

②

4 − 2 = _ _ _

Review: You can use counters to show subtraction. Take 3 counters away. Count the counters that are left: I. The subtraction equation is 4 minus 3 equals I.

Directions: I-2. How can you use counters to find the difference? Draw Xs on the counters to show the subtraction. Write the number to finish the subtraction equation.

Differentiation Resource Book

Subtract within 5

Name _____

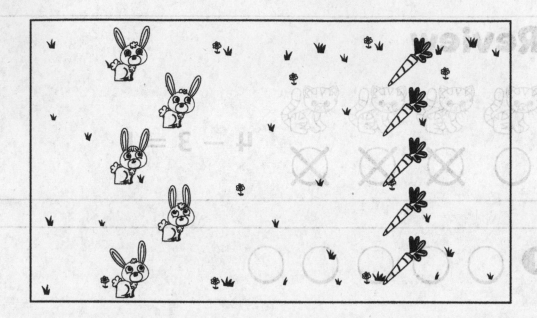

1 _____ _____ _____

‑ ‑ ‑ ‑ ‑ = ‑ ‑ ‑ ‑

_____ _____ _____

2 _____ _____ _____

‑ ‑ ‑ ‑ ‑ = ‑ ‑ ‑ ‑

_____ _____ _____

Directions: There are 5 rabbits and 5 carrots in a field. **1.** Tell a story about some of the rabbits hopping away. Draw Xs on the rabbits that hop away. Write a subtraction equation to match your story. **2.** Tell a story about some of the carrots being eaten. Draw Xs on the carrots that are eaten. Write a subtraction equation to match your story.

Differentiation Resource Book

Ways to Make 6 and 7

Name _____

Review

$5 + 1 = 6$

$1 + 5 = 6$

$4 + 2 = 6$

$2 + 4 = 6$

$3 + 3 = 6$

1 $6 + \_\_\_ = \_\_\_$

2 $5 + \_\_\_ = \_\_\_$

3 $\_\_\_ + \_\_\_ = \_\_\_$

Review: How can you make 6? 5 white counters and 1 black counter make 6, or 1 black counter and 5 white counters make 6. 4 white counters and 2 black counters make 6, or 2 black counters and 4 white counters make 6. 3 white counters and 3 black counters make 6.

Directions: How can you make 7? **1.** Color 1 counter. Write 1. Write the total. **2.** Color 2 counters. Write 2. Write the total. **3.** Color 3 counters. Write the numbers to complete an equation to match.

Differentiation Resource Book

Ways to Make 6 and 7

Name _____

①

②

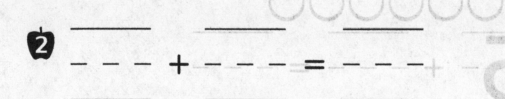

③

Directions: Louis sees apples in trees. **1.** How can you make 7? Draw a total of 7 apples in the trees. **2.** What equation matches your drawing? Write an equation to match. **3.** Write a different equation to match your drawing.

Ways to Decompose 6 and 7

Name _____

Review

$6 = 1 + 5$ $6 = 5 + 1$

$6 = 2 + 4$ $6 = 4 + 2$

$6 = 3 + 3$

1

$7 = \_\_\_ + \_\_\_$

2

$7 = \_\_\_ + \_\_\_$

3

$7 = \_\_\_ + \_\_\_$

Review: How can you decompose 6? 6 is 1 black triangle and 5 white triangles, or 6 is 5 white triangles and 1 black triangle. 6 is 2 black triangles and 4 white triangles, or 4 white triangles and 2 black triangles. 6 is 3 black triangles and 3 white triangles.

Directions: How can you decompose 7? **1.** Color 3 bears. Write 3. Count the white bears. Write the number. **2.** Color 5 bears. Write 5. Count the white bears. Write the number. **3.** Color 6 bears. Write 6. Count the white bears. Write the number.

Ways to Decompose 6 and 7

Name _____

1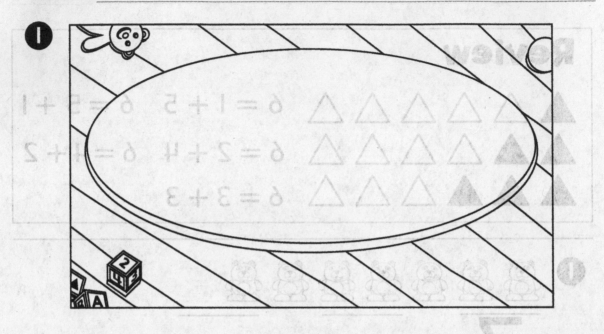

2 _____ = _____ + _____

3 _____ = _____ + _____

Directions: Mrs. Yun puts 6 books on the rug. **1.** How can she decompose 6? Draw 2 groups to show how she can break apart 6. **2.** What equation matches your drawing? Write an equation to match. **3.** Write a different equation to match your drawing.

Ways to Make 8 and 9

Name _____

Review

$$1 + 7 = 8 \quad 7 + 1 = 8$$
$$2 + 6 = 8 \quad 6 + 2 = 8$$
$$3 + 5 = 8 \quad 5 + 3 = 8$$
$$4 + 4 = 8$$

---

❶

$$2 + \text{---} = \text{---}$$

---

❷

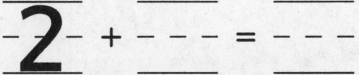

$$\text{---} + \text{---} = \text{---}$$

Review: How can you make 8? 1 black crayon and 7 white crayons make 8, or 7 white crayons and 1 black crayon make 8. 2 black crayons and 6 white crayons make 8, or 6 white crayons and 2 black crayons make 8. 3 black crayons and 5 white crayons make 8, or 5 white crayons and 3 black crayons make 8. 4 black crayons and 4 white crayons make 8.

Directions: How can you make 9? **1.** Color 2 paintbrushes. Count the white paintbrushes. Write the number. Write the total. **2.** Color 4 paintbrushes. Write 4. Count the white paintbrushes. Write the number. Write the total.

Differentiation Resource Book

Ways to Make 8 and 9

Name _____

1

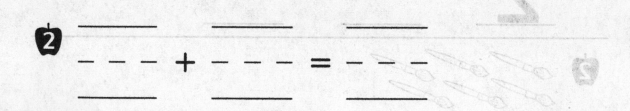

2 ___ ___ ___

— — — + — — — = — — —

___ ___ ___

3 ___ ___ ___ +

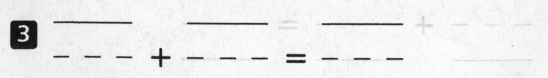

— — — + — — — = — — —

Directions: Hugo has 8 paintbrushes. **I.** How can Hugo make 8? Draw 2 groups that make 8. **2.** What equation matches your drawing? Write an equation to match. **3.** Circle paintbrushes to show another way to make 8. Write an equation to match.

Ways to Decompose 8 and 9

Name _____

Review

$9 = 1 + 8$ $9 = 8 + 1$
$9 = 2 + 7$ $9 = 7 + 2$
$9 = 3 + 6$ $9 = 6 + 3$
$9 = 4 + 5$ $9 = 5 + 4$

1

$8 = $ ___ $+$ ___

2

$8 = $ ___ $+$ ___

Review: How can you decompose 9? 9 is 1 black ball and 8 white balls, or 8 white balls and 1 black ball. 9 is 2 black balls and 7 white balls, or 7 white balls and 2 black balls. 9 is 3 black balls and 6 white balls, or 6 white balls and 3 black balls. 9 is 4 black balls and 5 white balls, or 5 white balls and 4 black balls.

Directions: How can you decompose 8? **1.** Color 3 balls. Write 3. Count the white balls. Write the number. **2.** Color 6 balls. Write 6. Count the white balls. Write the number.

Ways to Decompose 8 and 9

Name _____

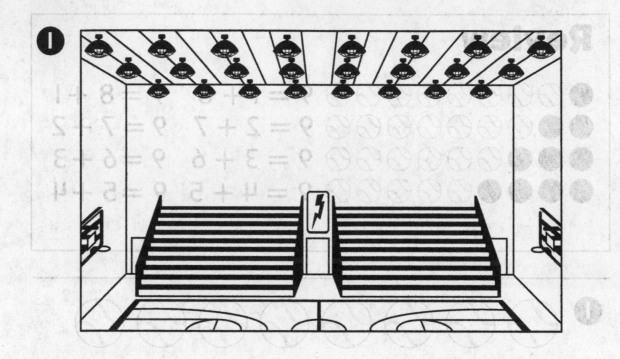

1

2 _____ = _____ + _____

3 _____ = _____ + _____

Directions: There are 9 fans in the stands. **1.** How can you decompose 9? Draw 2 groups of fans to show how you can break apart 9. **2.** What equation matches your drawing? Write an equation to match. **3.** Write a different equation to match your drawing.

Ways to Make 10

Name _____

Review

5 + 5 = 10

4 + 6 = 10   6 + 4 = 10

3 + 7 = 10   7 + 3 = 10

---

❶ ▢▢▢▢▢▢▢▢▢▢

 2 + _ _ _ = _ _ _

---

❷ ▢▢▢▢▢▢▢▢▢▢

 _ _ _ + _ _ _ = _ _ _

Review: How can you make 10? 5 star blocks and 5 moon blocks make 10. 4 star blocks and 6 moon blocks make 10, or 6 moon blocks and 4 star blocks make 10. 3 star blocks and 7 moon blocks make 10, or 7 moon blocks and 3 star blocks make 10.

Directions: What are other ways you can make 10? 1. Color 2 star blocks. Count the moon blocks, 1, 2, 3, 4, 5, 6, 7, 8. Write 8. 2 star blocks and 8 moon blocks, or 8 moon blocks and 2 star blocks make 10. Write 10. 2. Color 1 star block. Write 1. Count the moon blocks, 1, 2, 3, 4, 5, 6, 7, 8, 9. Write 9. 1 star block and 9 moon blocks, or 9 moon blocks and 1 star block make 10. Write 10.

Ways to Make 10

Name _____

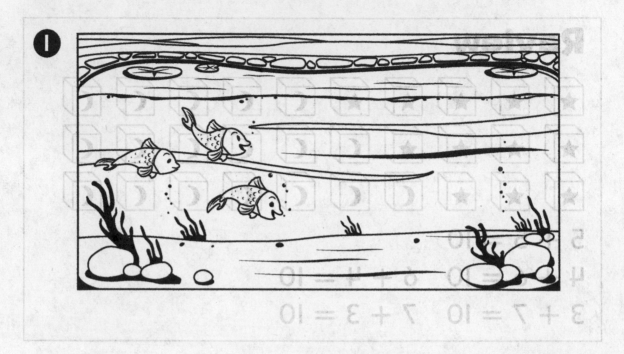

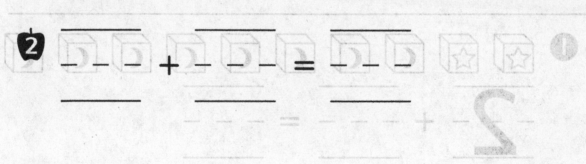

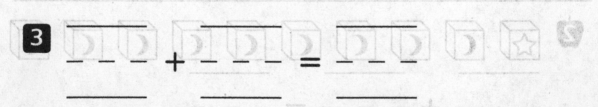

Directions: There are 10 fish in the pond. **1.** How can you make 10? Draw more fish to make 10. **2.** What equation matches your drawing? Write an equation to match. **3.** Draw a circle around a group of fish to show a different way to make 10. Write an equation to match your circled groups.

Ways to Decompose 10

Name _____

Review

$9 + 1 = 10$ $1 + 9 = 10$

$8 + 2 = 10$ $2 + 8 = 10$

$7 + 3 = 10$ $3 + 7 = 10$

❶

___ ___ = ___ + ___

❷

- - - = - - - + - - -

Review: How can you decompose 10? 10 is 9 black counters and 1 white counter, or 10 is 1 white counter and 9 black counters. 10 is 8 black counters and 2 white counters, or 10 is 2 white counters and 8 black counters. 10 is 7 black counters and 3 white counters, or 10 is 3 white counters and 7 black counters.

Directions: What are other ways you can decompose 10? **1.** Write 10. Color 6 counters red. Write 6. Count the white counters, 1, 2, 3, 4. Write 4. **2.** Write 10. Color 5 counters red. Write 5. Count the white counters, 1, 2, 3, 4, 5. Write 5.

Ways to Decompose 10

Name _____

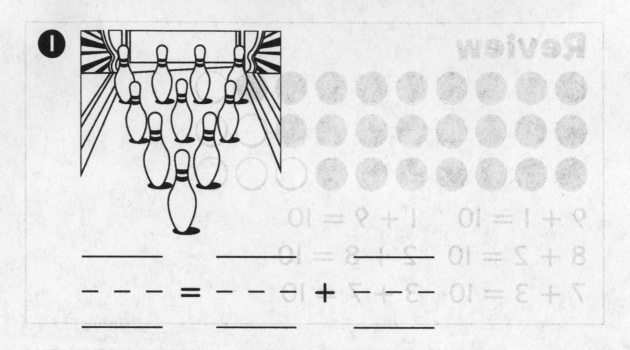

①

_____   _____

_ _ _ _ = _ _ _ _ + _ _ _ _

_____   _____

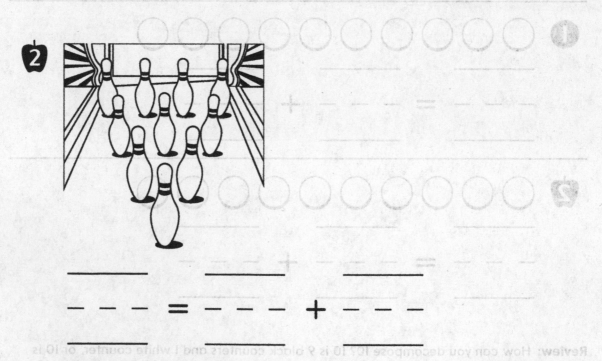

②

_____   _____   _____

_ _ _ _ = _ _ _ _ + _ _ _ _

_____   _____   _____

Directions: Tay and Jax are bowling. **1.** Tay bowls first. Circle a group of pins to show which pins he knocks down. Write an equation to match. **2.** Jax bowls next. Circle a group of pins to show which pins he knocks down. Write an equation to match.

Differentiation Resource Book

Lesson 9-1 • Reinforce Understanding

Represent 11, 12, and 13

Name _____

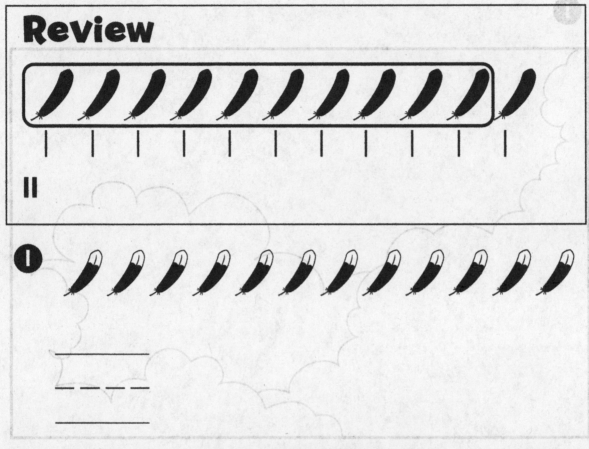

Review How many feathers are there? Draw a line below each feather as you count. When you count 10 feathers, circle a group of 10, 1, 2, 3, 4, 5, 6, 7, 8, 9, 10. Draw a line below each feather as you count the feather that is not circled, 1. There are 11 feathers.

Directions: 1–2. How many feathers are there? Draw a line below each feather as you count. When you count 10, circle a group of 10. Then keep counting the feathers that are not circled. Trace the number to show how many feathers are in the group.

Represent 11, 12, and 13

Name

①

②
_____
_ _ _ _
_____

Directions: Sarah sees 11 birds in the sky. 1. Draw 11 birds in the sky. Circle 10 birds. Explain how you know that you drew 11 birds. **2.** What number shows the number of birds in your drawing? Write the number that shows how many birds are in your drawing.

Differentiation Resource Book
98

Make 11, 12, and 13

Name _____

Review

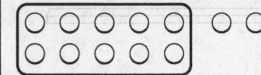

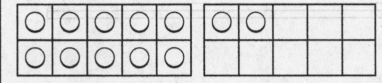

$$10 + 2 = 12$$

❶
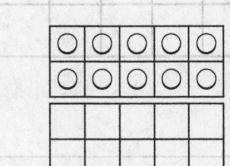

$$10 + \text{---} = \text{-----}$$

❷
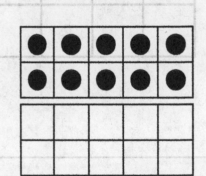

$$10 + \text{---} = \text{-----}$$

Review: How can you make 12? Circle a group of 10. You can put 10 counters in a ten-frame. Then you can put 2 more counters in another ten-frame. The equation to match this is 10 plus 2 equals 12.

Directions: 1. How can you make 11? Circle a group of 10. Draw the counter that is not circled in the empty ten-frame. Write the number of counters you drew in the equation. Trace 11. **2.** How can you make 13? Circle a group of 10. Draw the counters that are not circled in the empty ten-frame. Write the number of counters you drew in the equation. Trace 13.

Differentiation Resource Book

Make 11, 12, and 13

Name _____

①

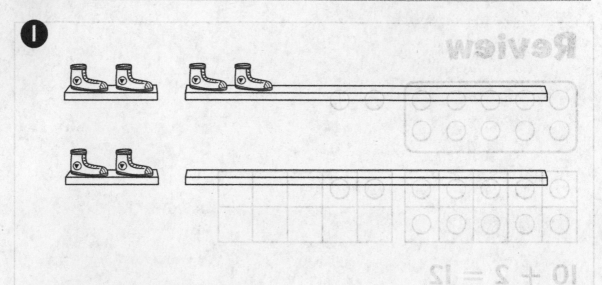

②

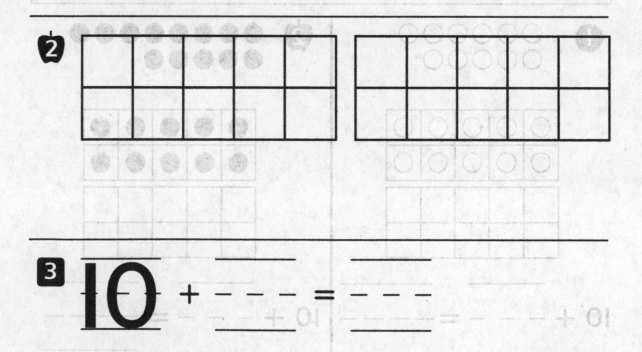

③ $10 + \text{----} = \text{----}$

Directions: Niko has 12 shoes on shelves at his shoe store. 1. How can you draw more shoes to make 12? Draw more shoes to make 12. Explain how you know how many shoes to draw. **2.** How can you draw counters to make 12? Draw counters in the ten-frames to show how you can make 12. **3.** Write an equation for your ten-frames.

Decompose 11, 12, and 13

Name _____

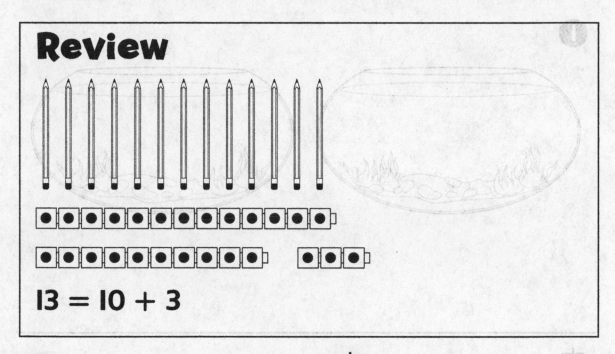

Review

$$13 = 10 + 3$$

❶

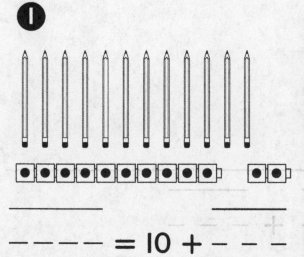

___ ___
___ ___ = 10 + ___

❷

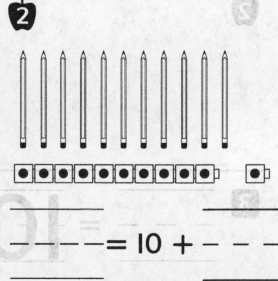

___ ___
___ ___ = 10 + ___

Review: How can you decompose 13? Match 1 cube to each pencil. Break off 10 connecting cubes. Count the cubes that are left, 1, 2, 3. 13 equals 10 plus 3.

Directions: 1. How can you decompose 12? Trace 12. Circle a group of 10 cubes. Count the extra cubes that are not circled. Write the number of extra cubes you counted. **2.** How can you decompose 11? Trace 11. Circle a group of 10 cubes. Count the extra cubes that are not circled. Write the number of extra cubes you counted.

Differentiation Resource Book

Decompose 11, 12, and 13

Name _____

①

②

③

$$\_ \_ \_ = 10 + \_ \_ \_$$

Directions: A pet store has 13 fish. 1. How can you decompose 13? Draw 10 fish in one bowl. How can you draw fish in the other bowl to show 13 in all? Draw fish in the other bowl to show 13 in all. Explain how you know how many fish to draw. **2.** How can you draw cubes to show how you decomposed 13? Draw cubes to show how you decomposed 13. **3.** Write an equation to match your cubes.

Differentiation Resource Book

102

Represent 14 and 15

Name _____

Review

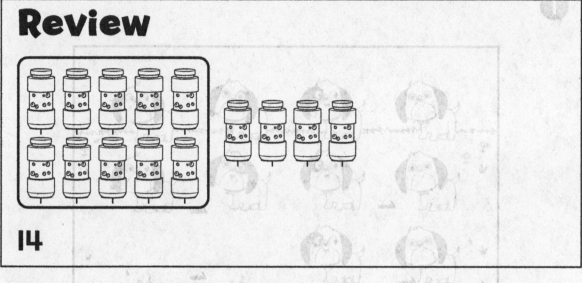

14

❶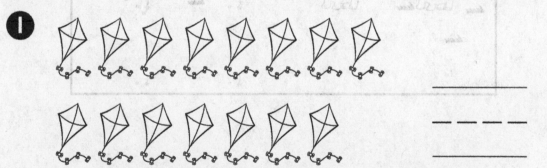

_____

- - - - -

_____

❷

_____

- - - - -

_____

Review How many bottles are there? Draw a line on each bottle as you count. When you count to 10 bottles, circle a group of 10. Count the bottles that are not circled, 11, 12, 13, 14. There are 14 bottles.

Directions: 1. How many kites are there? Draw a line on each kite as you count. Circle a group of 10. Then keep counting the kites that are not circled. Trace the number to show how many kites are in the group. **2.** How many jacks are there? Draw a line on each jack as you count. Circle a group of 10. Then keep counting the jacks that are not circled. Trace the number to show how many jacks are in the group.

Differentiation Resource Book

Represent 14 and 15

Name _____

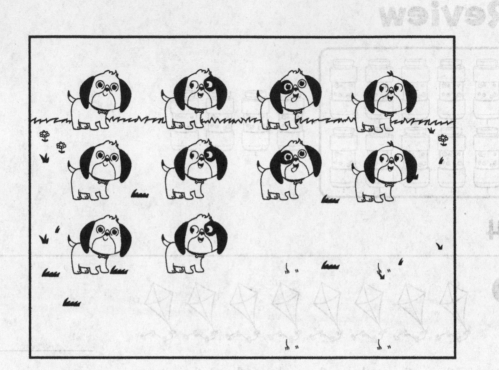

_____

- - - - - - - - -

_____

Directions: 14 dogs are at the park. 1. How can you draw more dogs to show 14? Draw more dogs to show 14. Explain your thinking. **2.** Write the number that shows how many dogs.

Differentiation Resource Book
104

Lesson 9-5 • Reinforce Understanding

Make 14 and 15

Name _____

Review

10 + 5 = 15

1

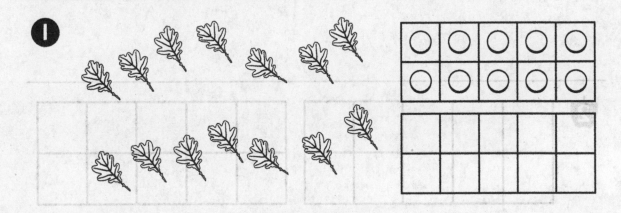

_____   _____

10 + – – – – = – – – –

_____   _____

Review: How can you make 15? Circle a group of 10 flowers. Put I counter for each flower in the ten-frame. Count the rest of the flowers. Put I counter for each of the rest of the flowers in the second ten-frame. There is a group of ten and a group of some more. The equation 10 plus 5 equals 15 shows the number of flowers.

Directions: I. How can you make 14? Circle a group of 10. Draw I counter in the empty ten-frame for each leaf that is not circled. Write the number of counters you drew in the equation. Trace 14.

Differentiation Resource Book

105

Copyright © McGraw-Hill Education

Make 14 and 15

Name _____

1

2

 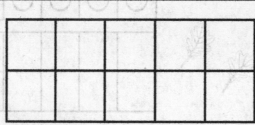 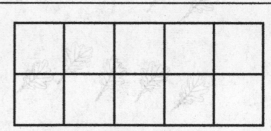

3 $\mathbf{10}$ + _ _ _ _ = _ _ _ _

Directions: There are 15 ears of corn growing in a field. **1.** How can you draw more ears of corn to make 15? Draw more ears of corn to show 15. Explain your thinking. **2.** How can you draw counters in the ten-frames to show the corn in the field? Draw counters in the ten-frames to show the corn in the field. **3.** Write an equation to match your ten-frames.

Differentiation Resource Book

106

Decompose 14 and 15

Name _____

Review

$$14 = 10 + 4$$

1

___ ___ = **10** + ___

Review: How can you decompose 14? Match 1 cube to each pear. Break off 10 connecting cubes. Count the cubes that are left, 1, 2, 3, 4. 14 equals 10 plus 4.

Directions: 1. How can you decompose 15? Trace 15. Circle a group of 10 cubes. Count the cubes that are not circled. Write the number of extra cubes you counted.

Decompose 14 and 15

Name _____

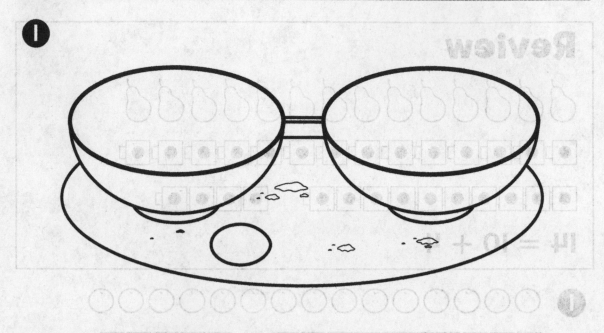

1

2

$$_\,_\,_ = 10 + _\,_\,_$$

3

$$_\,_\,_\,_ = 10 + _\,_\,_$$

Directions: José helps his grandmother gather eggs. He has 14 eggs. The eggs will go in 2 bowls. 1. How could José decompose 14 eggs? Draw eggs in the bowls to show your thinking. 2. How can you draw cubes to show the eggs in the bowls? Draw cubes to represent the eggs. 3. Write the equation to match your cubes.

Differentiation Resource Book

108

Represent 16 and 17

Name _____

Review

---

❶

_____

- - - - -

_____

❷

_____

- - - - -

_____

Review: How many butterflies? Count the butterflies. Count each butterfly only one time: 1, 2, 3, 4, 5, 6, 7, 8, 9, 10, 11, 12, 13, 14, 15, 16, 17. There are 17 butterflies. **Directions: 1-2.** How many flowers? Count the flowers. Write the number that shows how many flowers.

Represent 16 and 17

Name _____

❶

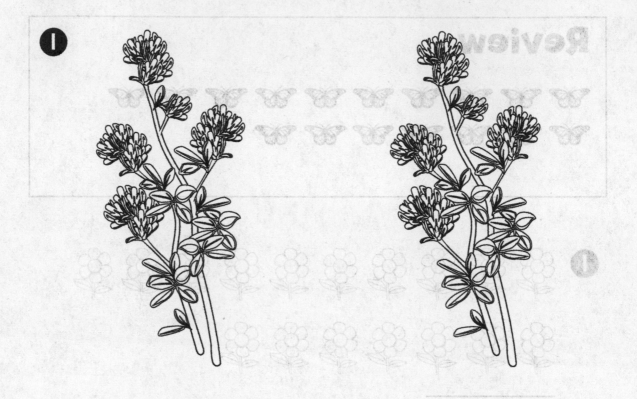

❷

Directions: 1. Draw 16 butterflies in or around the flowers. **2.** Draw 17 butterflies in or around the trees.

Make 16 and 17

Name _____

Review

$$10 + 7 = 17$$

1 $\quad 10 + \underline{} = 16$

2 $\quad 10 + \underline{} = 17$

Review: How can you make 17? Seventeen is ten ones and seven more ones. **Directions:**
1. What is the missing number? Write the missing number to make 16. 2. What is the missing number? Write the missing number to make 17.

Make 16 and 17

Name _____

Directions: 1. Sixteen is ten ones and some more ones. Color ten ducks green. Color some more ducks blue to make sixteen. **2.** Seventeen is ten ones and some more ones. Color ten birds red. Color some more birds orange to make seventeen.

Decompose 16 and 17

Name _____

Review

$$17 = 10 + 7$$

① $16 = \underline{\quad\quad} + 6$

② $17 = \underline{\quad\quad} + 7$

Review: How can you decompose 17? Seventeen can be broken down into ten ones and seven ones.
Directions: 1. What is the missing number? Write the missing number to decompose 16.
2. What is the missing number? Write the missing number to decompose 17.

Decompose 16 and 17

Name _____

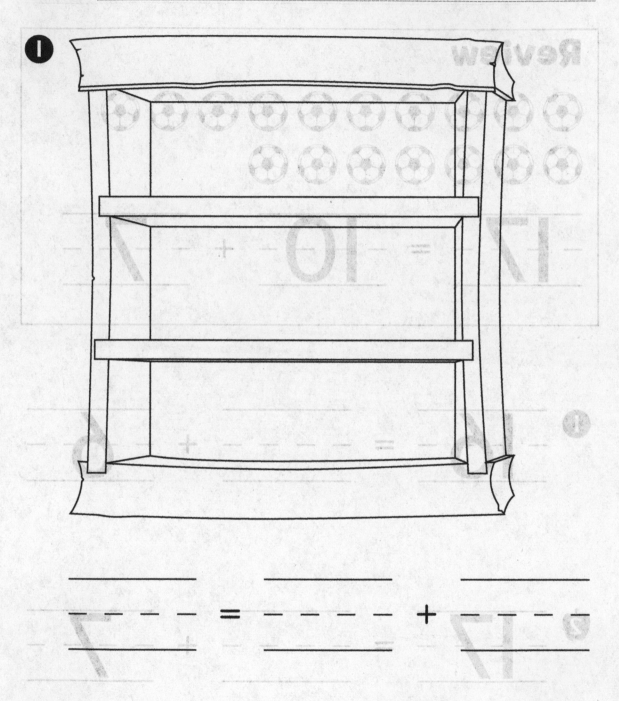

①

$$\underline{\quad} \, \underline{\quad} = \underline{\quad} \, \underline{\quad} + \underline{\quad} \, \underline{\quad}$$

Directions: 1. Oliver has 17 toys to put on the shelves. He knows he wants to put ten toys on one shelf. How many toys would he put on the second shelf? Draw the toys on the shelves. Write the equation to show 17.

Represent 18 and 19

Name _____

Review

(1)

[keys]

1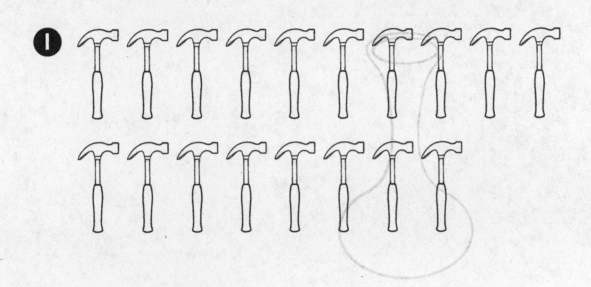

_____
_ _ _ _
_____

Review: How many keys? Count the keys. Count each key only one time: 1, 2, 3, 4, 5, 6, 7, 8, 9, 10, 11, 12, 13, 14, 15, 16, 17, 18, 19. There are 19 keys. **Directions: 1.** How many hammers? Count the hammers. Write the number that shows how many hammers.

Represent 18 and 19

Name _____

1

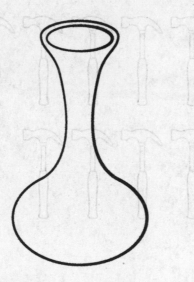

Make 18 and 19

Name _____

Review

$$10 + 9 = 19$$

1. $10 + \underline{} = 18$

2. $10 + \underline{} = 19$

Review: How can you make 19? Nineteen is ten ones and nine more ones. **Directions:**
1. What is the missing number? Write the missing number to make 18. **2.** What is the
missing number?
Write the missing number to make 19.

Make 18 and 19

Name _____

1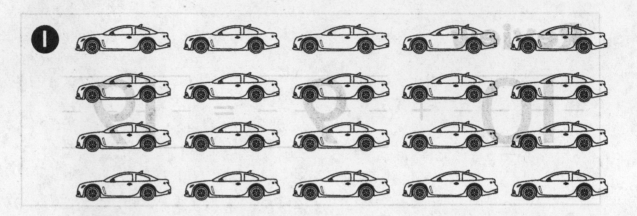

2

Directions: 1. Eighteen is ten ones and some more ones. Color ten cars purple. Color some more cars orange to make eighteen. **2.** Nineteen is ten ones and some more ones. Color ten trucks yellow. Color some more trucks pink to make nineteen.

Differentiation Resource Book

118

Decompose 18 and 19

Name _____

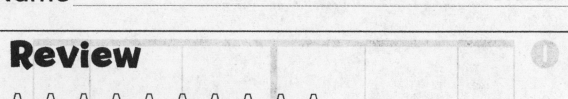

Review

$$19 = 10 + 9$$

① $19 = \underline{\qquad} + 9$

② $18 = \underline{\qquad} + 8$

Review: How can you decompose 19? Nineteen can be broken down into ten ones and nine ones.

Directions: 1. What is the missing number? Write the missing number to decompose 19.
2. What is the missing number? Write the missing number to decompose 18.

Lesson 10-6 • Extend Thinking

Decompose 18 and 19

Name

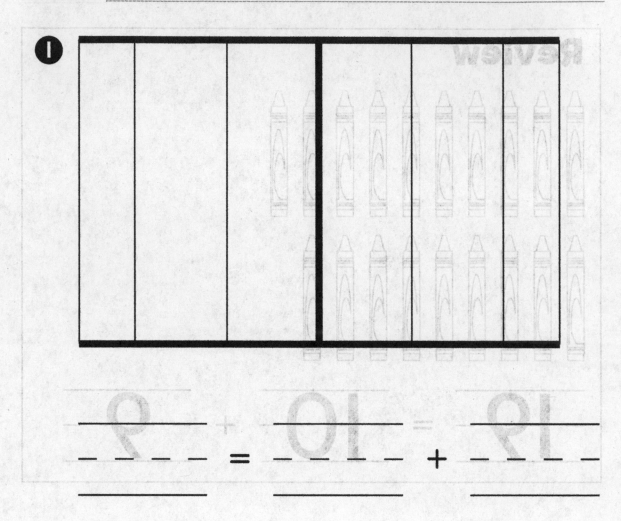

Directions: 1. Some players are at football practice. There are ten players on one side. There are nine players on the other side. How many players are there in all? Draw the players on the field. Write the equation.

Differentiation Resource Book

120

2-Dimensional and 3-Dimensional Shapes

Name _____

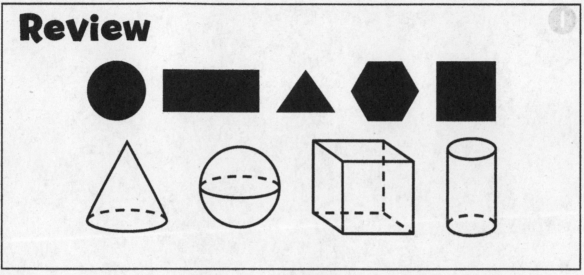

Review

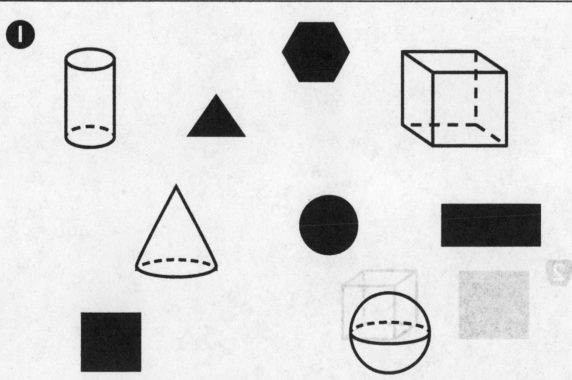

Review: Are these 2-dimensional shapes or 3-dimensional shapes? The 2-dimensional shapes are flat. The 3-dimensional shapes are solid. **Directions: 1.** Which shapes are 2-dimensional? Which shapes are 3-dimensional? Put an X on the 2-dimensional shapes. Circle the 3-dimensional shapes.

Differentiation Resource Book
121

2-Dimensional and 3-Dimensional Shapes

Name _____

1

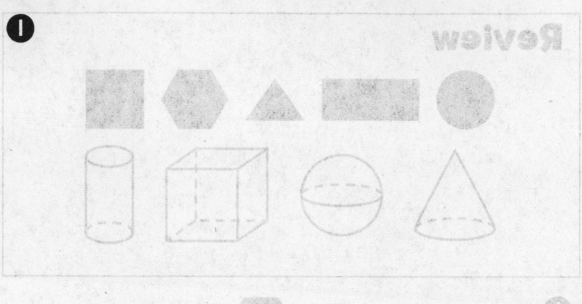

2

Directions: 1. How can you draw three 2-dimensional shapes? Describe each shape.
2. Which shape is 3-dimensional? Circle the 3-dimensional shape. Describe the shape.

Differentiation Resource Book

Cubes

Name _____

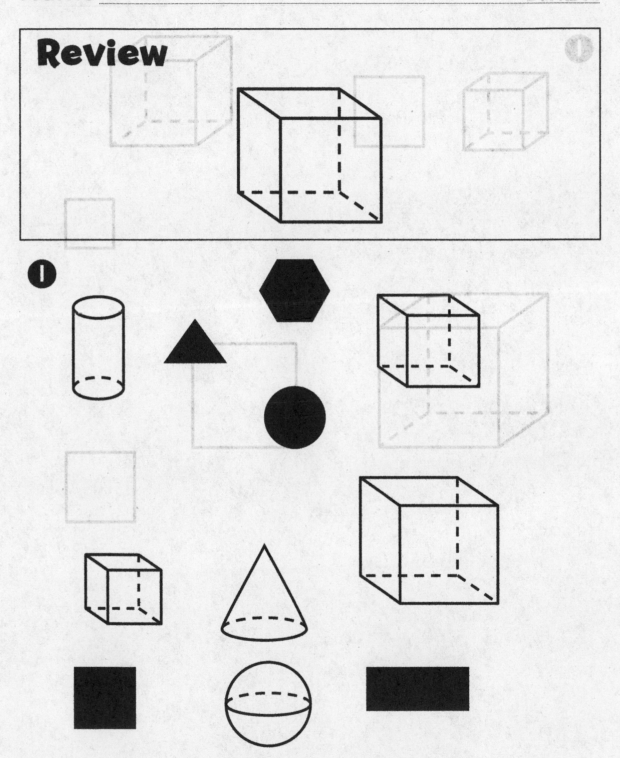

Review: What 3-dimensional shape is this? This is a cube.

Directions: 1. Which of these shapes are cubes? Circle all the cubes.

Differentiation Resource Book

Cubes

Name _____

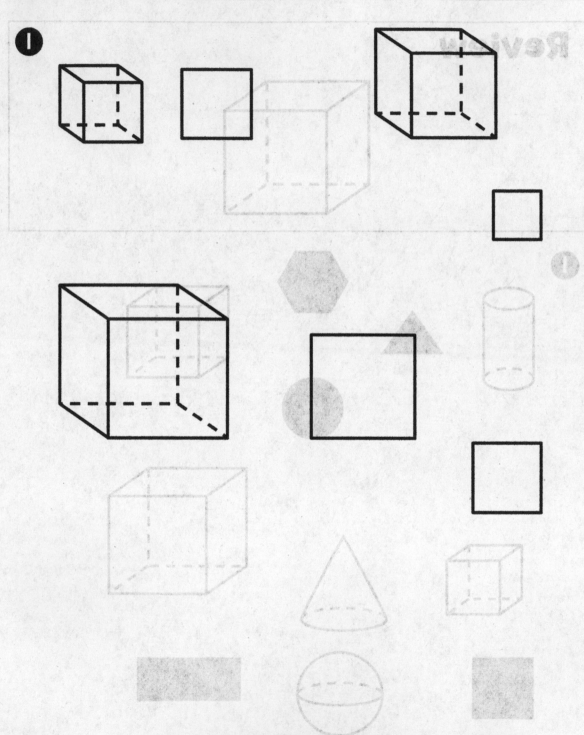

❶

Directions: 1. How would you compare a square and a cube? A square sometimes looks like a cube, but a square is flat. Color the cubes green. Color the squares red

Spheres

Name _____

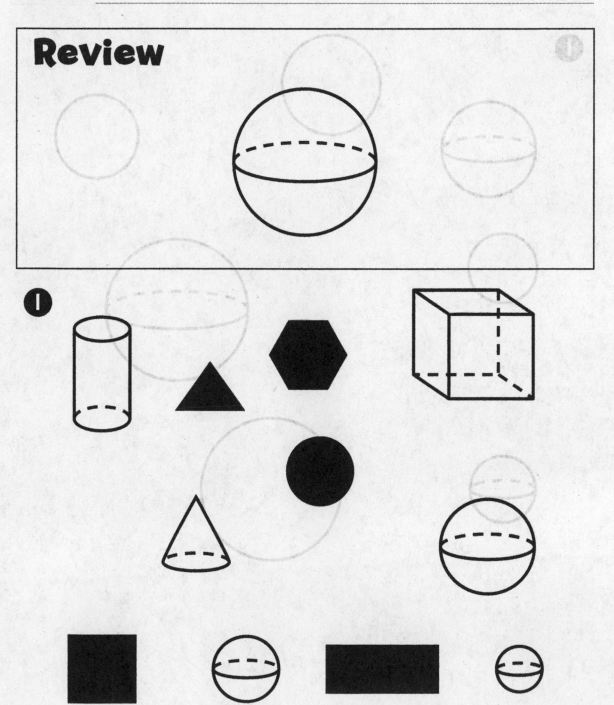

Review: What 3-dimensional shape is this? This is a sphere.
Directions: I. Which of these shapes are spheres? Circle all the spheres.

Lesson II-3 • Extend Thinking

Spheres

Name _____

Lesson II-3 • Reinforce Und

Spheres

Name

1

Directions: I. How would you compare a circle and a sphere? A circle sometimes looks like a sphere, but a circle is flat. Color the spheres blue. Color the circles yellow.

Lesson 11-4 • Reinforce Understanding

Cylinders

Name _____

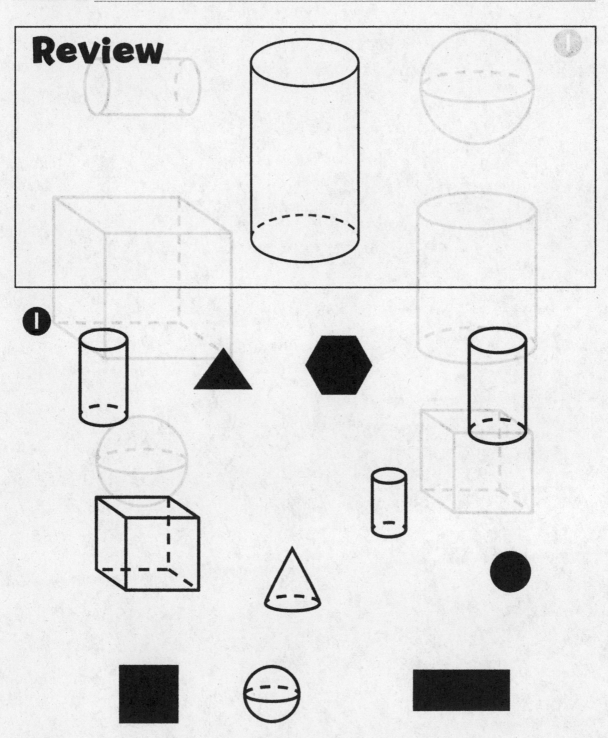

Review: What 3-dimensional shape is this? This is a cylinder.

Directions: 1. Which of these shapes are cylinders? Circle all the cylinders.

Cylinders

Name _____

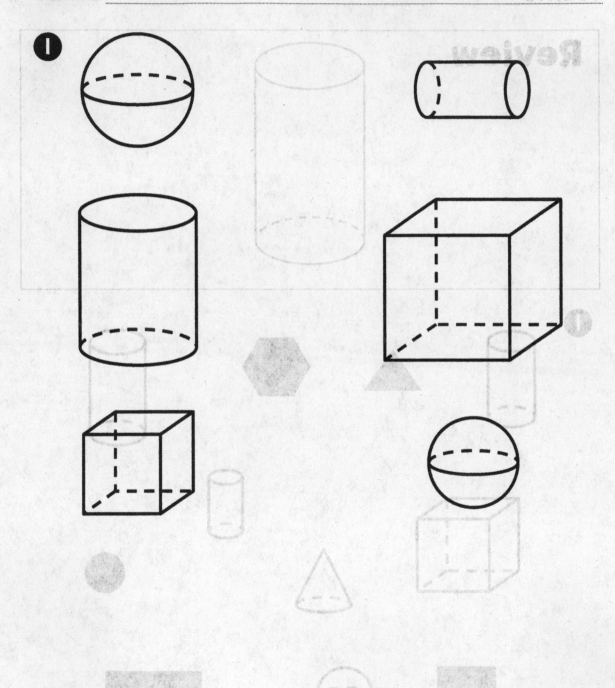

Directions: 1. Which shapes can roll? Which shapes cannot roll? A cylinder can roll. Color the shapes that can roll purple. Color the shapes that cannot roll yellow.

Differentiation Resource Book

Cones

Name _____

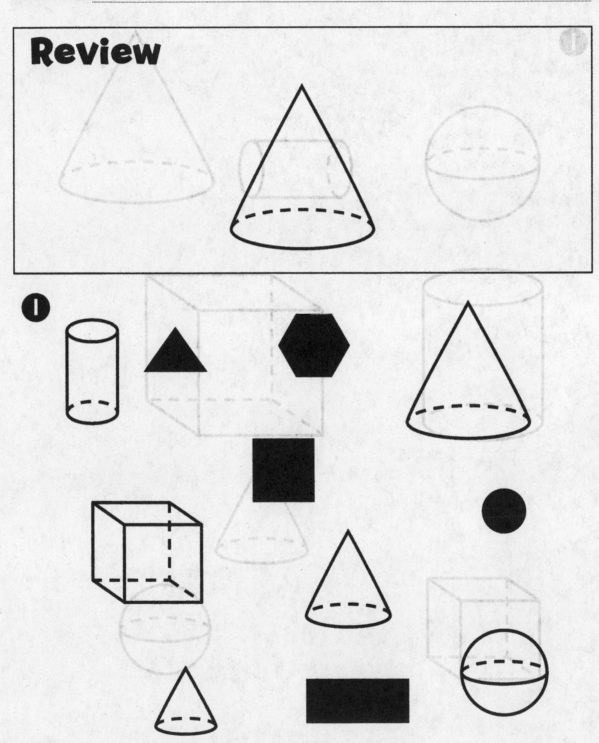

Review: What 3-dimensional shape is this? This is a cone.
Directions: 1. Which of these shapes are cones? Circle all the cones.

Cones

Name

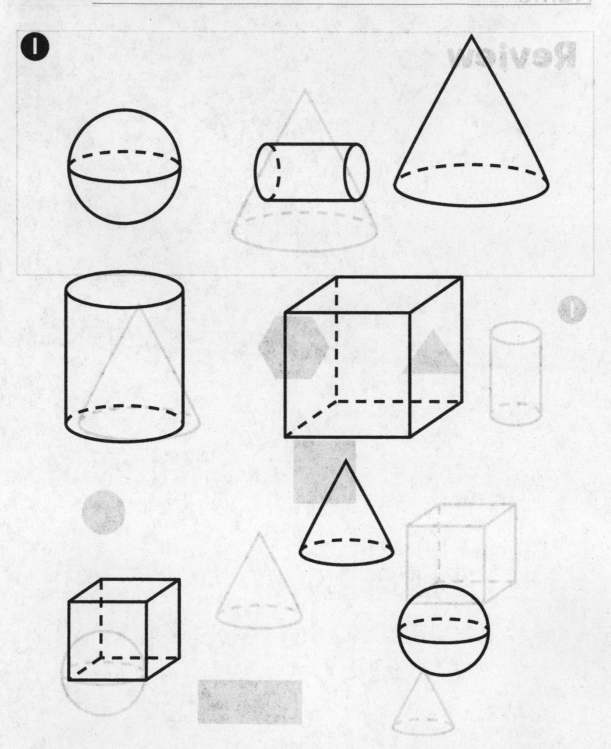

Directions: 1. Which shapes can roll? Which shapes cannot roll? A cone can roll. Color the shapes that can roll red. Color the shapes that cannot roll purple.

Differentiation Resource Book
130

Lesson 11-6 • Reinforce Understanding
Describe Solids

Name _____

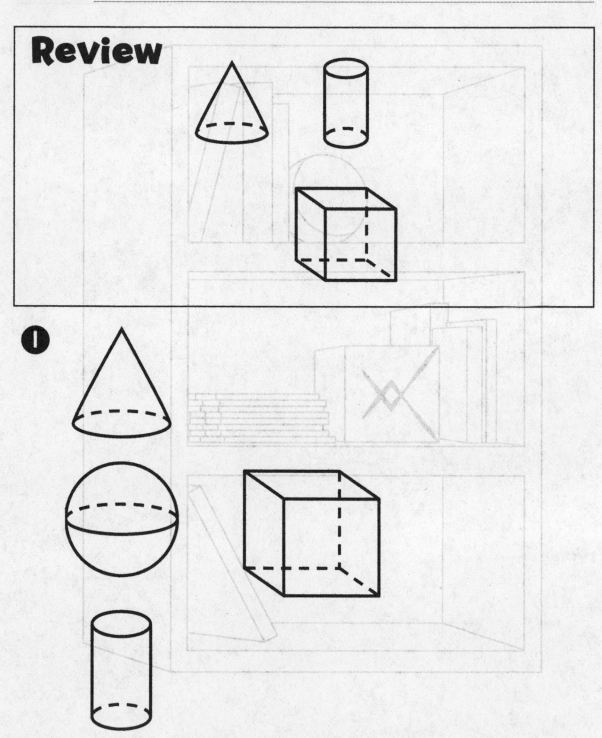

Review: How can you describe these 3-dimensional shapes? The cone is beside the cylinder. The cube is below the cylinder.

Directions: 1. Which shape is beside the sphere? Put an X on the shape that is beside the sphere.

Differentiation Resource Book
131

Describe Solids

Name _____

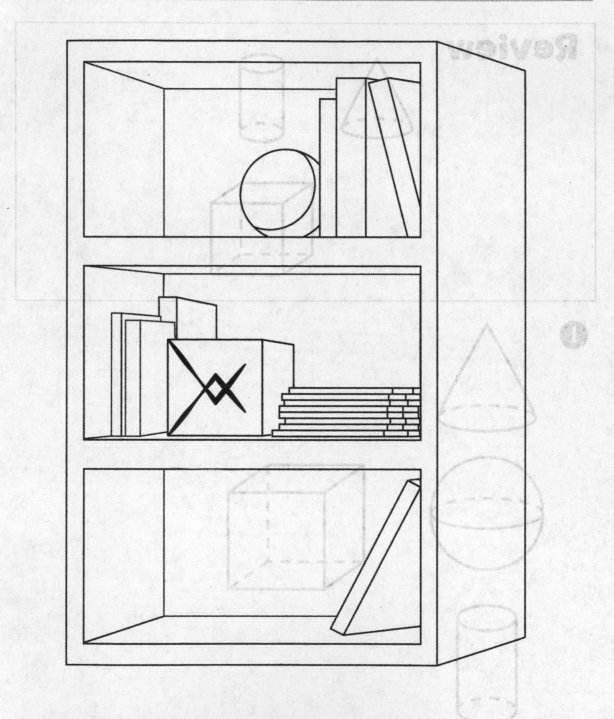

Directions: How can you describe the location of 3-dimensional shapes? Draw a sphere that looks like a ball below the book on the bottom shelf. Draw a cone that looks like a party hat above the box on the middle shelf. Draw a cube that looks like an aquarium next to the ball on the top shelf. Describe the objects you drew using 3-dimensional names.

Count by 1s to 50

Name _____

Review

1	2	3	4	5	6	7	8	9	10
11	12	13	14	15	16	17	18	19	20
21	22	23	24	25	26	27	28	29	30
31	32	33	34	35	36	37	38	39	40
41	42	43	44	45	46	47	48	49	50

❶

1	2	3	4	5	6	7	8	9	10
11	12	13	14	15	16	17	18	19	20
21	22	23	24	25	26	27	28	29	30
31	32	33	34	35	36	37	38	39	40
41	42	43	44	45	46	47	48	49	50

Review: Nathan has toy horses. How many toy horses does he have? You can use the number chart to help you count the horses. Start at 1. Count 1, 2, 3, and so on. Point to each horse as you count. When you get to the end of a row, move down to the next row. Start at the left end of the next row when you count. Nathan has 50 toy horses.

Directions 1: Point to each number as you count. Which numbers are missing? Trace the missing numbers as you count.

Differentiation Resource Book

Count by 1s to 50

Name _____

①

1	2	3	4	5	6	7	8	9	10
11	12	13	14	15	16	17	18	19	20
21	22	23	24	25	26	27	28	29	30
31	32	33	34	35	36	37	38	39	40
41	42	43	44	45	46	47	48	49	50

_____

②

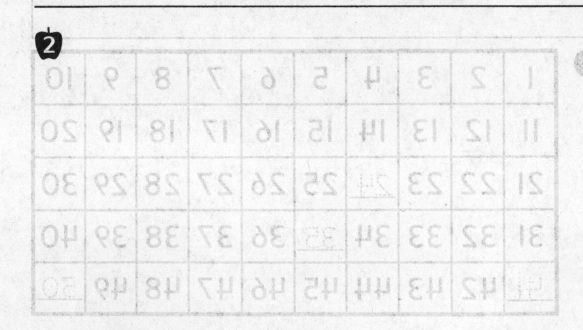

Directions: Vin plays a missing number game. 1. What numbers are missing? Write the numbers that are missing from the chart. **2.** What patterns do you see on the chart? Use colors to show patterns on the chart. Explain the pattern you used.

Count by 1s to 100

Name _____

Review

1	2	3	4	5	6	7	8	9	10
11	12	13	14	15	16	17	18	19	20
21	22	23	24	25	26	27	28	29	30
31	32	33	34	35	36	37	38	39	40
41	42	43	44	45	46	47	48	49	50
51	52	53	54	55	56	57	58	59	60
61	62	63	64	65	66	67	68	69	70
71	72	73	74	75	76	77	78	79	80
81	82	83	84	85	86	87	88	89	90
91	92	93	94	95	96	97	98	99	100

Review: How do you count by 1s to 100? You can use counters to help you count. Put one counter on each number as you count. Look for patterns in the table.

Directions: Which numbers are missing? Trace the numbers to complete the chart while you count.

Count by 1s to 100

Name _____

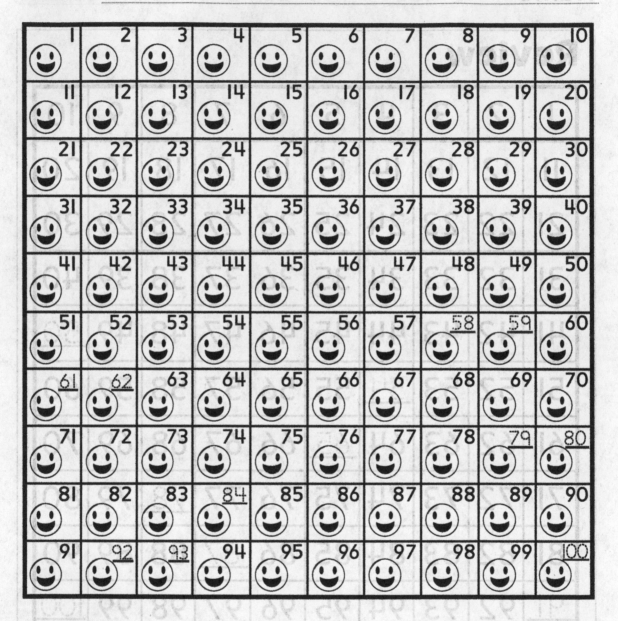

Directions: Rahi counts stickers. Help her count the stickers and complete the chart. 1. What numbers are missing? Write the missing numbers. 2. Color sticker number 59 red. Color sticker number 62 green. Color sticker number 79 orange. Color sticker number 84 purple. Color sticker number 100 blue.

Lesson 12-3 • Reinforce Understanding

Count by 10s to 100

Name _____

Review

○○○○○○○○○○ 10
○○○○○○○○○○ 20
○○○○○○○○○○ 30
○○○○○○○○○○ 40
○○○○○○○○○○ 50
○○○○○○○○○○ 60
○○○○○○○○○○ 70
○○○○○○○○○○ 80
○○○○○○○○○○ 90
○○○○○○○○○○ 100

1

1	2	3	4	5	6	7	8	9	10
11	12	13	14	15	16	17	18	19	20
21	22	23	24	25	26	27	28	29	30
31	32	33	34	35	36	37	38	39	40
41	42	43	44	45	46	47	48	49	50
51	52	53	54	55	56	57	58	59	60
61	62	63	64	65	66	67	68	69	70
71	72	73	74	75	76	77	78	79	80
81	82	83	84	85	86	87	88	89	90
91	92	93	94	95	96	97	98	99	100

Review: How do you count by 10s to 100? You can use cubes to help you count. Count the cubes in one train: 1, 2, 3, 4, 5, 6, 7, 8, 9, 10. There are 10 cubes in each train. Count by 10s: 10, 20, 30, 40, 50, 60, 70, 80, 90, 100.

Directions 1: How can you count by 10s? Trace the numbers to complete the chart while you count.

Differentiation Resource Book
137

Copyright © McGraw-Hill Education

Count by 10s to 100

Name _____

1 10
20
30
40
50
60
70
80
90
100

Directions: Jiro has marbles in rows of 10. How many marbles does Jiro have? **1.** How many marbles are there? Count the marbles by 10s. Write the numbers as you count. Explain how you use patterns to count.

Differentiation Resource Book

138

Count From Any Number to 100

Name _____

Review

41	42	43	**44**	45	46	47	48	49	50
51	52	53	54	55	56	57	58	59	60

①

11	12	13	14	15	16	17	18	19	20
21	**22**								
		33	34	35	36	37	38	39	40
41	42	43	44	45	46	47	48	49	50
51	52	53	54	55	56	57	58	59	60
61	62	63	64	65	66	**67**			
							78	79	80

②

Review: Start at 44. What numbers come next when you count by 1s? The next numbers are 45, 46, 47, 48, 49, 50, 51, 52, and so on.

Directions: 1. Use the number chart. Start at 22. What numbers come next when you count by 1s? Count the numbers by 1s. Color the numbers as you count. **2.** Start at 67. What numbers come next when you count by 1s? Count the numbers by 1s. Color the numbers in a different color as you count.

Differentiation Resource Book

Count From Any Number to 100

Name _____

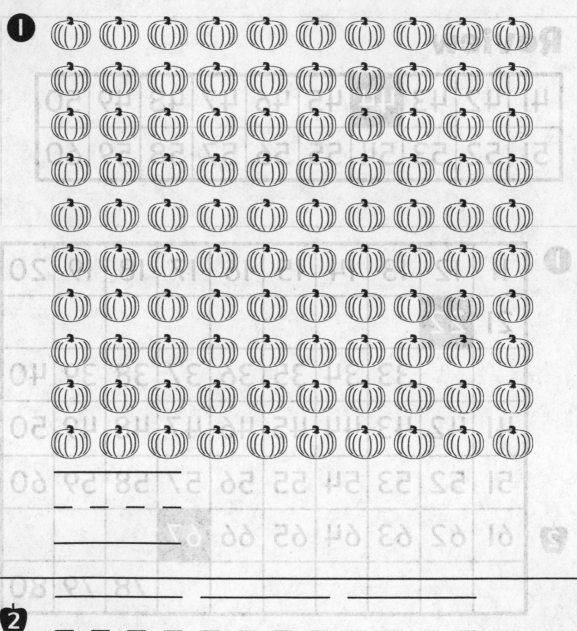

1

2

___ ___ ___ ___ ___ ___ ___

- - - - - - - - - - -

___ ___ , ___ ___

Directions: Duane sees pumpkins on the farm. 1. Choose some number of pumpkins. Write your number. Color that number of pumpkins orange. **2.** What numbers come next when you count by 1s? Write the numbers that come next when you count by 1s.

Count to Find Out How Many

Name _____

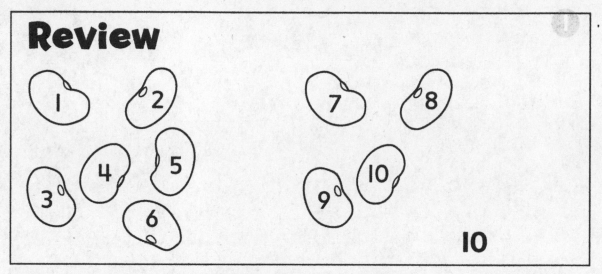

Review

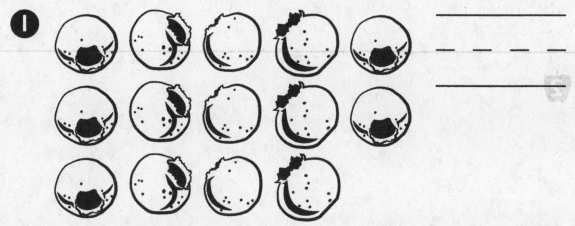

10

①

②

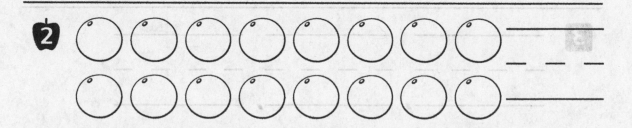

Review: How many beans? Count the number in the first group: 1, 2, 3, 4, 5, 6. There are 6 beans in the first group. Count the beans in the second group, starting with the number that comes after 6: 7, 8, 9, 10. There are 10 beans.

Directions 1–2: How many? Count. Write the numbers on each fruit to help you count. Write the number.

Count to Find Out How Many

Name _____

1

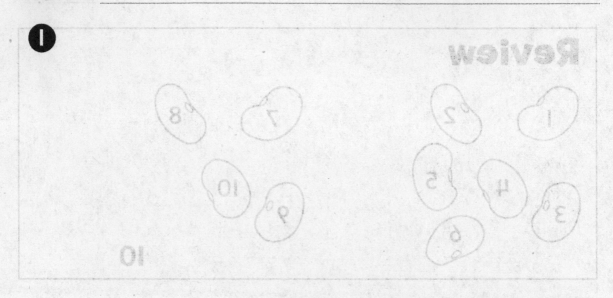

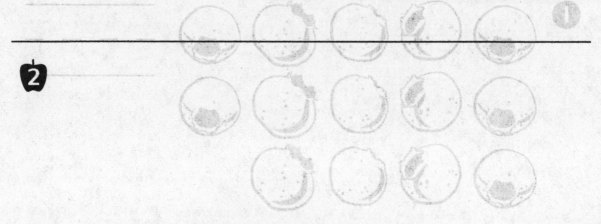

2

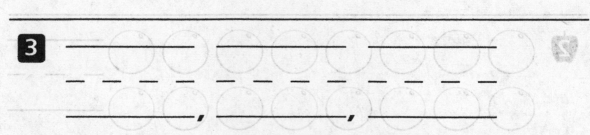

3

Directions: Isa has I7 coins. 1. How can you draw I7 coins? **2.** How can you show I7 coins in another way? **3.** What numbers come after I7 when you count by Is? Write the numbers that come next.

Compare and Contrast 2-Dimensional Shapes

Name _____

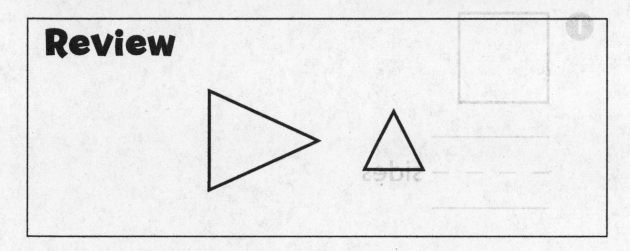

Review

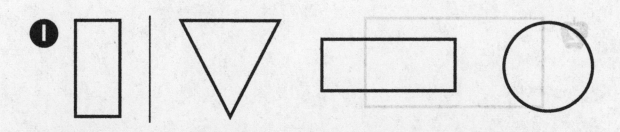

❶

❷

Review: How are these 2-dimensional shapes the same? They are both triangles. They both have 3 sides. They both have 3 vertices.

Directions 1–2. How can you compare the shapes? Circle the shape that is the same as the first shape.

Compare and Contrast
2-Dimensional Shapes

Name _____

❶

_____

− − − − − sides

_____

❷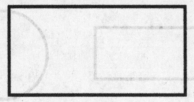

_____

− − − − − vertices

_____

3

Directions: 1. How many sides does this shape have? **2.** How many vertices does this shape have? **3.** Circle the shape that has the same number of sides and vertices as the shapes above.

Lesson 13-2 • Reinforce Understanding
Build and Draw
2-Dimensional Shapes

Name

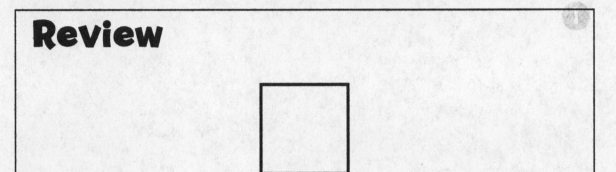

Review

1

2

Review: What do you know about this 2-dimensional shape? This is a square. It has 4 sides. All 4 sides are the same length.

Directions 1. Draw a 2-dimensional shape that has 3 sides and 3 vertices. **2.** Draw a 2-dimensional shape that has 0 sides and 0 vertices.

Build and Draw
2-Dimensional Shapes

Name _____

1

_____

– – – – sides

_____

2

_____

– – – – sides

_____

Directions: 1. Draw a shape that has 3 vertices. Write how many sides it has. **2.** Draw a shape with 1 more side than the shape above. Write how many sides it has.

Compose 2-Dimensional Shapes

Name

Review

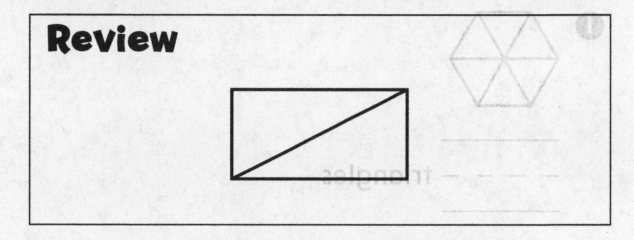

❶

❷

Review: What shapes is this 2-dimensional shape made of? This rectangle is made of 2 triangles.

Directions 1. Circle the shape that is made of 3 triangles. **2.** Circle the shape that is made of 4 squares.

Compose 2-Dimensional Shapes

Name _____

1

_____

– – – – – triangles

_____

2

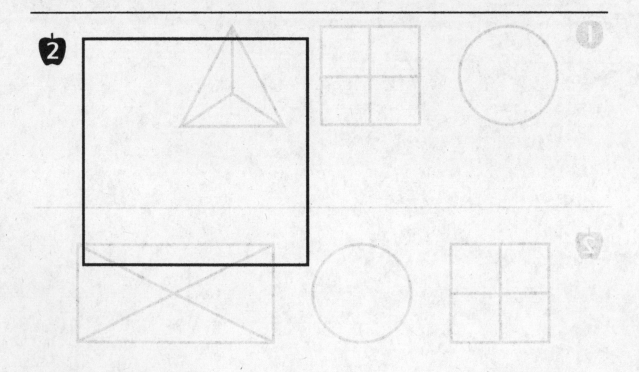

Directions: 1. How many triangles is this shape made of? **2.** Draw lines to show smaller squares in this large square.

Lesson 13-4 • Reinforce Understanding
Compare and Contrast 3-Dimensional Shapes

Name _____

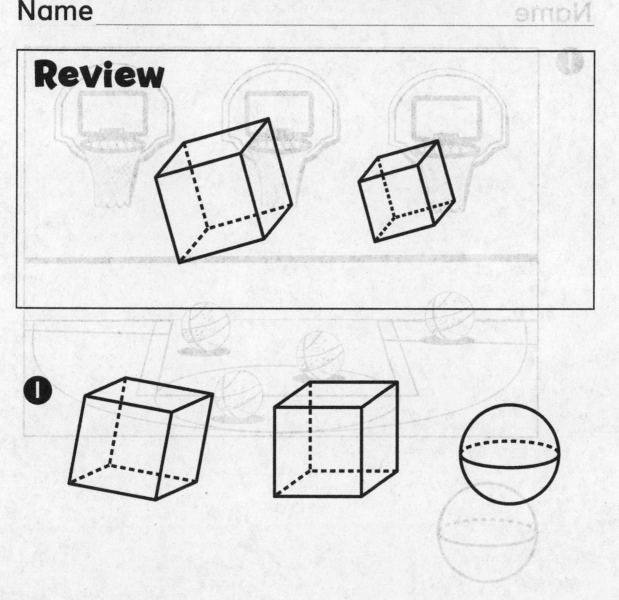

Review: How are these shapes the same? Both of them have 6 flat faces.

Directions: 1. How can you compare the shapes? Put an X on the shape that is different.

Compare and Contrast 3-Dimensional Shapes

Name _____

1

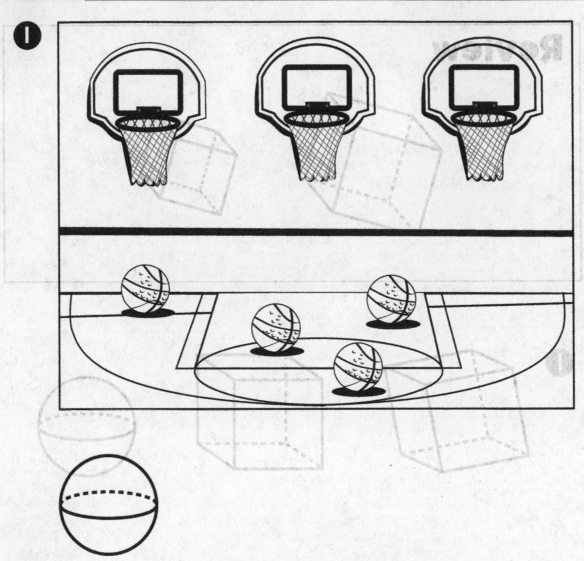

Directions: 1. Circle the objects in the picture that are the same as the 3-dimensional object shown.

Build 3-Dimensional Shapes

Name _____

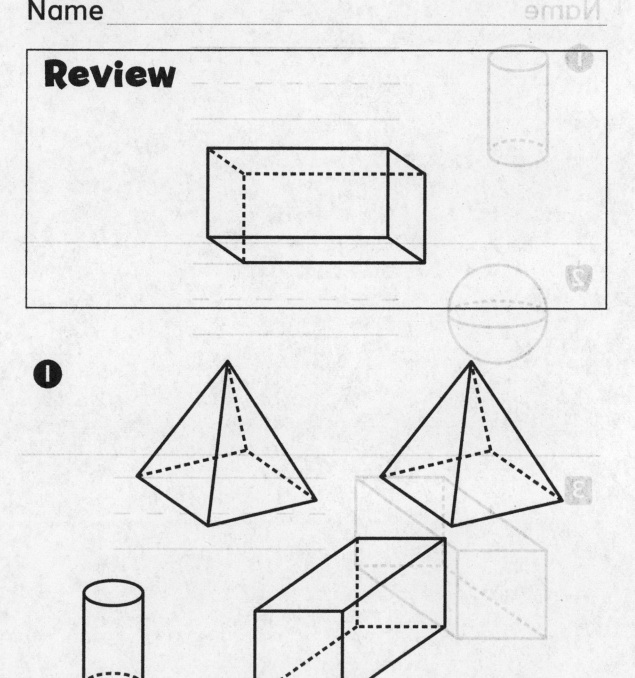

Review

I

Review: What do you know about this shape? It has flat faces. It is a solid shape.

Directions: I. How can you make these shapes? Put an X on the shape that you could roll clay to make.

Build 3-Dimensional Shapes

Name _____

1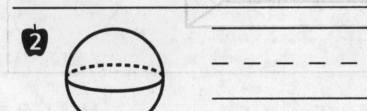

2

3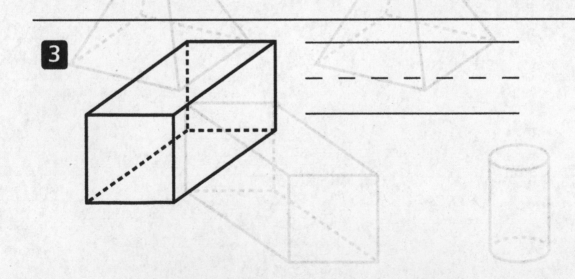

Directions: 1. If you made this shape with clay would it roll? **2.** If you made this shape with clay would it stack? **3.** If you made this shape with clay would it roll?

Describe 3-Dimensional Shapes in the World

Name

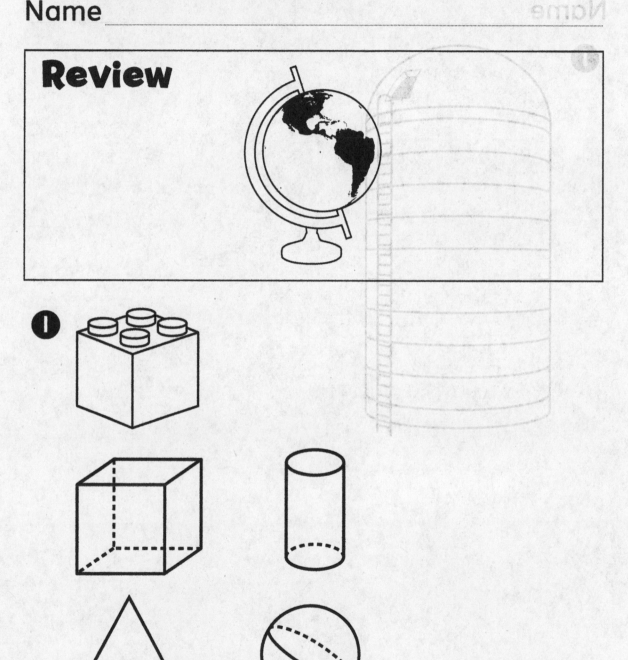

Review

①

Review: What 3-dimensional shape does this globe look like? The main part is a sphere.

Directions: I. What shape does the building block look like? Circle the 3-dimensional object that the block looks like.

Describe 3-Dimensional Shapes in the World

Name

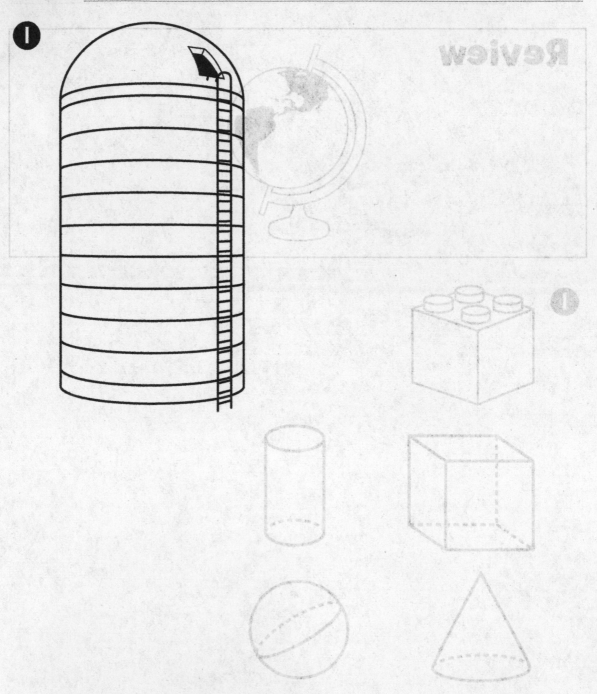

Directions: I. What shapes is this silo made of? Circle the cylinder. Put an X on the half sphere.

Describe Attributes of Objects

Name _____

Review

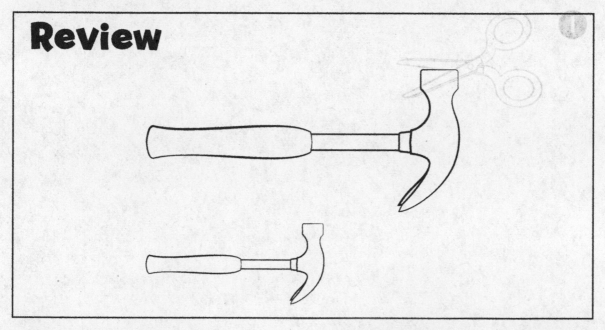

①

②

Review: How can we describe the objects? We can describe the hammers by talking about how long they are. We can also describe the objects by how heavy they are.

Directions: 1–2. Circle the object that you can describe by its capacity.

Describe Attributes of Objects

Name _____

①

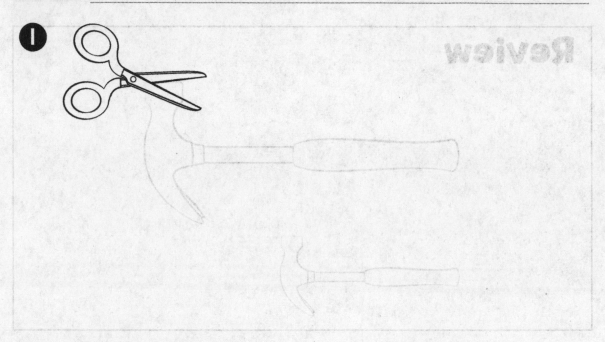

②

Directions: 1. What is heavier? What is longer? Draw something that is heavier and longer than scissors. **2.** What is shorter? What is lighter? Draw something that is shorter and lighter than a barn.

Compare Lengths

Name _____

Review

1

2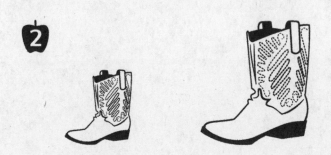

Review: How can we compare these objects by length? The whale is longer than the fish.

Directions: 1. Which object is longer? Circle the object that is longer. **2.** Which object is shorter? Put an X on the object that is shorter.

Compare Lengths

Name _____

1

Directions: 1. Draw something that is longer than the toothbrush. Then draw something that is even longer.

Compare Heights

Name _____

Review

1

Review: How can we compare these objects by height? The guitar is taller than the drum.

Directions: 1. Which object is taller? Circle the object that is taller.

Compare Heights

Name _____

1

Directions: 1. Draw something that is shorter than the tree. Then draw something that is even shorter.

Lesson 14-4 • Reinforce Understanding

Compare Weights

Name _____

Lesson 14-4 • Exte

Compare W

Name

Review

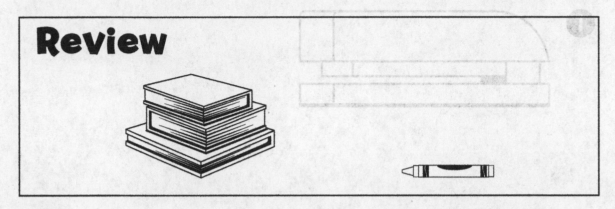

1

2

Review: How can we compare these objects by weight? The books are heavier than the crayon.

Directions: 1. Which is heavier? Circle the object that is heavier. **2.** Which is lighter? Put an X on the object that is lighter.

Compare Weights

Name

1

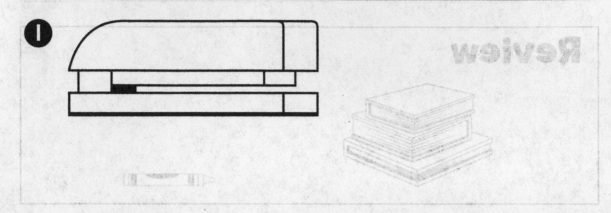

Directions: 1. Draw something that is heavier than the stapler. Then draw something that is even heavier.

Lesson 14-5 • Reinforce Understanding
Compare Capacities
Name

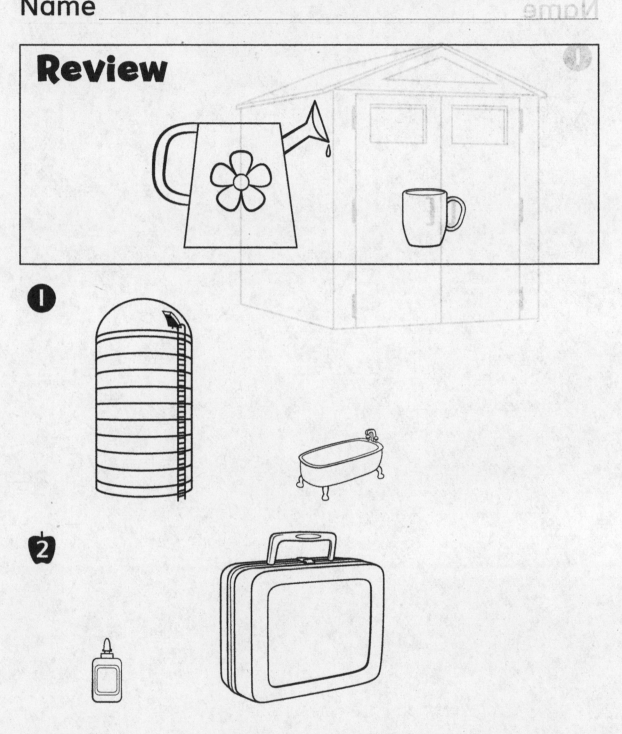

Review: How can we compare these objects by capacity? The watering can holds more than the mug.

Directions: I. Which holds more? Circle the object that holds more. **2.** Which holds less? Put an X on the object that holds less.

Lesson 14-5 • Extend Thinking

Compare Capacities

Name _____

①

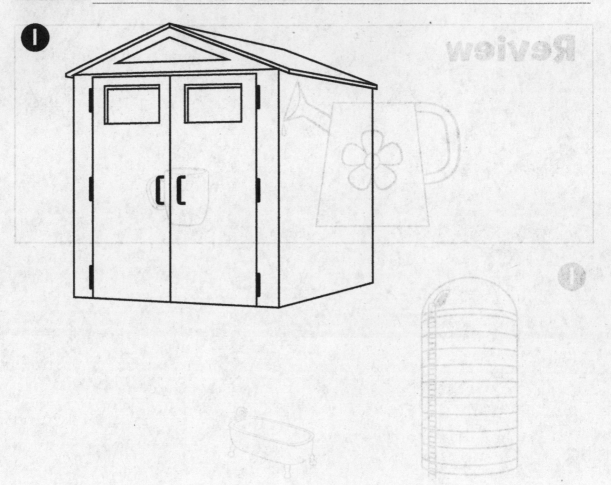

Directions: I. Draw something that holds less than the shed. Then draw something that holds even less.

Differentiation Resource Book

164